Theory and Practices on Innovating
for Sustainable Development

Yoram Krozer

Theory and Practices on Innovating for Sustainable Development

 Springer

Yoram Krozer
Department of Governance and Technology
  for Sustainability (CSTM)
University of Twente
Enschede, The Netherlands

The Sustainable Innovations Academy
Amsterdam, The Netherlands

ISBN 978-3-319-18635-1   ISBN 978-3-319-18636-8   (eBook)
DOI 10.1007/978-3-319-18636-8

Library of Congress Control Number: 2015942918

Springer Cham Heidelberg New York Dordrecht London
© Springer International Publishing Switzerland 2016
This work is subject to copyright. All rights are reserved by the Publisher, whether the whole or part of the material is concerned, specifically the rights of translation, reprinting, reuse of illustrations, recitation, broadcasting, reproduction on microfilms or in any other physical way, and transmission or information storage and retrieval, electronic adaptation, computer software, or by similar or dissimilar methodology now known or hereafter developed.
The use of general descriptive names, registered names, trademarks, service marks, etc. in this publication does not imply, even in the absence of a specific statement, that such names are exempt from the relevant protective laws and regulations and therefore free for general use.
The publisher, the authors and the editors are safe to assume that the advice and information in this book are believed to be true and accurate at the date of publication. Neither the publisher nor the authors or the editors give a warranty, express or implied, with respect to the material contained herein or for any errors or omissions that may have been made.

Printed on acid-free paper

Springer International Publishing AG Switzerland is part of Springer Science+Business Media (www.springer.com)

*This book is dedicated to my father,
Simon Krozer.*

Preface

This book *Theory and Practices on Innovating for Sustainable Development* is about various ways in pursuing income along with better environmental qualities. It presents the state of knowledge and shows possibilities and impediments of innovating in energy, water, tourism, consumption, transport, and culture with the aim of sustainable development. It also delivers proposals about how to go forward. The diversity of people and their capabilities enable innovations for sustainable development. There are many ways to develop and use products and services with large social benefits and to resolve impediments for such activities.

The idea to write a book about innovations for sustainable development was born at the Cartesius Institute, Institute for Sustainable Innovations of the Netherlands Technical Universities. During 15 years after the start in 1989, it has created education on energy and environment management; made studies on product development for renewable energy, water, and land uses; and supported firms pursuing sustainable innovations. Hundreds of students and scholars of the technical universities from all over the world cooperated with inventors, entrepreneurs, authorities, and social organizations. For the book, these experiences are enriched with the economic theories and own practices in business and social organizations. The Cartesius Institute derailed a year after change of management.

Many people contributed to the insights, particularly hundreds of scholars and inventors involved in the Institute. It is not possible to mention all. I do want to mention the cofounders Elisabeth (Lies) van der Ven (scholar) and Anita Andriessen (politician) who are not with us anymore, as well as protectors during stormy periods André Olijslager (entrepreneur), Luuk Hermans and Simon Tijsma (policy), and Han Brezet (scholar). Numerous young innovators were involved thanks to Siem Jansen, Pieter Smit, Wini Weidenaar, and Andries van Weperen (entrepreneurs); Bouwe de Boer, Ferd Crone, and Ed Nijpels (policy); Thea Bijma, Satish Kumar Beela, Hilde van Meerendonk, and Sharon Tokich-Hophmayer (scholars), as well as Wubbe Ockels (astronaut and scholar) who passed away. Regional culture and nature as resources for innovations are an insight indebted to Femke van den Akker; Begoña Angulo Urien; Jeni Bujini; Alison Mcinnes; Albert Ruiter; Eric Vos; Jannewietske de Vries; Jerzy Wcisła (policy); Mohammed Al Taani; Else

Christensen-Redzepovic; Oleg Dyakov; Nynke-Rixt Jukema; Tim Laning; Leo de Kok; Igor Studennikov; Jan Tichelaar; Jan-Kees Vis; Tinus Vos; Pauline Westerdorp; Anne-Jan Zwart, as well as Stephan Jansen and Erik Meijer who passed away (entrepreneurs); Shyam Asolekar; Hans Bressers; Frans Coenen; Marcel Crul; Donald Huisingh; Rahul Kamble; Ina Macaione; Ezio Manzini; Michael Narodoslawsky; Chris Ryan; and Ursula Tischner (scholars).

I am grateful to professor Andries Nentjes for the discussions and comments he provided; it is encouraging when Nestor of Environmental Economics in Europe ends his critiques with: "Interesting. No big comments anymore." Mira Krozer and Freek Willems comment on the readability of the book, and their help with editing is highly appreciated. I am grateful to former students of the Cartesius Institute Martha Chadyiwa, Alicia Sweder-Foster, and Zubeida Zwavel for correcting my imperfect English and comments.

Enschede, The NetherlandsYoram Krozer
Amsterdam, The Netherlands

Contents

1 Introduction 1
   1.1   Goals and Terms 1
   1.2   Environmental Debate 5
   1.3   Innovating for Environment 8
   1.4   Content of the Book 14
   References 15

2 Income Growth and Environmental Impacts 19
   2.1   IPAT Model 19
   2.2   Decoupling Impacts and Income 21
   2.3   Conventional Explanations 25
         2.3.1   Natural Resource Prices 25
         2.3.2   'Outsourcing' Environmental Impacts 26
         2.3.3   Environmental Policies 27
   2.4   Innovation and Decoupling 28
   2.5   Conclusions 30
   Appendix 31
   References 33

3 Markets of Sustainable Innovations 35
   3.1   Inducing Innovations 35
   3.2   Demands for Sustainable Innovations 36
   3.3   Markets of Sustainable Innovations 38
         3.3.1   Material Resources 39
         3.3.2   Pollution Control 39
         3.3.3   Ethical Consumption 40
         3.3.4   Ecosystem Services 41
         3.3.5   Cultural Attributes 41
         3.3.6   Summary 42
   3.4   Impediments for Sustainable Innovations 42
   3.5   Conclusions 45
   References 46

4	**User Innovations in Solar Power**	49
	4.1 Innovating Users	49
	4.2 Markets for Solar Boats	52
	4.3 Users' Perspective	53
	4.4 Market Perspective	55
	4.5 Conclusion	57
	References	57
5	**Tacit Inventors in Regions**	59
	5.1 Clusters and Networks	59
	5.2 Policy on Tourism	61
	5.3 Inventors in the Network Policy	63
	5.4 Consortia in the Cluster Policy	65
	5.5 Conclusions	69
	Appendix	69
	References	75
6	**Arts for Environmental Qualities**	77
	6.1 Arts and Services	77
	6.2 Arts Services Performance	79
	6.3 Arts Service of Natural Blends	83
	6.4 Conclusions	85
	References	85
7	**Technology Suppliers for Sanitation**	87
	7.1 Maxi–Min Regulation	87
	7.2 Conventional Sanitation	89
	7.3 Sanitation Alternatives	91
	7.3.1 Separation Network	91
	7.3.2 Distributed Network	92
	7.3.3 Value-Adding Services	95
	7.4 Conclusions	96
	References	97
8	**Alternatives for Commuting**	99
	8.1 Real Estate Cycles	99
	8.2 Mitigation of Congestion	101
	8.3 Alternative Offices	103
	8.4 Life Cycle Costs of Offices	105
	8.5 Conclusions	107
	Appendix	108
	References	110
9	**Ethical Consumers and Producers**	113
	9.1 Consumers' Will	113
	9.2 Ethical Consumption	115

Contents   xi

	9.3	Ethical Supplies	117
	9.4	Decision Making	119
	9.5	Conclusion	121
	References		122
10	**Sustainable Investors and Innovators**		**125**
	10.1	Innovation Process	125
	10.2	Investing in Innovations	127
	10.3	Policy Support of Innovators	128
	10.4	Investors' and Innovators' Interests	130
	10.5	Conclusions	135
	References		136
11	**Energy Services for Smart Grid**		**137**
	11.1	Local Energy Initiatives	137
	11.2	Energy Service Companies	139
	11.3	Support of Energy Consumption	143
	11.4	Support of Energy Production	145
	11.5	Conclusions	148
	Appendix		149
	References		158
12	**Renewable Energy Business and Policy**		**161**
	12.1	Renewable Energy Business	161
	12.2	Barriers and Drivers	163
	12.3	Policies and Renewable Energy Business	167
	12.4	Regional Policies	171
	12.5	Conclusions	173
	Appendix		174
	References		179
13	**Conclusions on Sustainable Innovations**		**183**
	13.1	Innovations for Sustainable Development	183
	13.2	Fostering Diversity of Sustainable Innovators	185
		13.2.1 Innovating Consumers	185
		13.2.2 Tacit Inventors	186
		13.2.3 Arts Services	186
		13.2.4 Technology Suppliers	187
		13.2.5 Office Alternatives	188
		13.2.6 Ethical Consumption	188
		13.2.7 Investors in Sustainable Innovations	189
		13.2.8 Energy Markets	190
		13.2.9 Renewable Energy Business	190
	13.3	On Sustainable Development	191

Index **193**

Chapter 1
Introduction

How to create income and foster a good environment? This question is the subject of the book based on cooperation with innovators pursuing for sustainable development. This subject is presented in the context of economic development and environmental qualities evolving throughout the last century and culminating in the notion of sustainable development. The innovations, it means doing things differently, are considered as being the results of the know-how interactions aiming at a benefit. The diversity of innovators is underscored. Next to research and development organisations in institutions and corporations, one finds technology suppliers, small- and medium-scale companies, individual tinkerers, artists and designers, non-governmental organisations and policymakers that innovate due to exceptional skills. The mainstream neoclassic theory with its focus on prices, the evolutionary theory about knowledge-related processes and the behavioural theory that addresses social behaviour are discussed with regard to policies. More policy or market is not necessarily the solution because more individual and social demands for good environment could do right things.

1.1 Goals and Terms

The natural environment with its various biological (living) and physical qualities such as materials, energy, space, nature and so on is continuously transformed into resources for human activities. The mass and space stay after these transformations on Earth but are often degraded, fragmented and dispersed and at best recoverable for some uses. The transformations of environmental qualities are purposed to create welfare. It means satisfying individual and social demands and aspirations through generating and distributing of wealth, leisure, care and other values as well as enabling decision making about these values across generations, sexes and races, all these given scarce resources. Private and social incomes, herewith, are essential tools. Hence, income growth and fair distribution are generally demanded

(Sen 2009). From the mainstream economic perspective on welfare, three kinds of resources are involved: capital that is indicated by income and assets; labour indicated by jobs, human relations, income, individual and social development, personal satisfaction and so on and environmental qualities or natural resources indicated by materials, energy, pollution, land use, nature and suchlike (Stiglitz et al. 2010). How income can grow and environmental qualities be improved is the subject of this book. Good labour conditions and fair distribution of income and wealth are considered preconditions for the welfare growth. These issues are largely neglected. They are comprehensively covered in other works (e.g. Jackson 2011; Piketty 2013).

In the mainstream, neoclassic argumentation, incomes grow when capital, labour and environmental qualities are allocated efficiently, which means when the lowest costs are achieved given prices of these resources. It is often assumed that capital is spontaneously allocated in an efficient manner, which is a peculiar viewpoint with regard to the recurrent financial collapses throughout centuries (Ferguson 2008). Economists usually consider allocation of labour. The environmental economists add allocation of environmental qualities. Given work hours per person, labour resources enlarge when populations grow. The efficient allocation of the growing labour is supposed to increase productivity. It means higher output to input measured in the monetary terms. Given allocation of capital and environmental resources, a higher labour productivity entails higher gross income. This income embraces wages, rents and profits. The sum of the individual income in an area or country is expressed as the gross regional or national income. The gross national income equals by definition the total national expenditure or the gross national production (GNP). The latter equals the gross domestic product (GDP) plus the imported minus exported income; the global GDP equals global gross income. People usually aim at more gross income and policies at the GDP growth though these are deficient indicators of welfare if uncorrected for inflation to assess the real income, for purchasing power to compare countries and social classes and for the social costs of activities that undermine welfare, such as degradation of the environment. Herewith, labour and capital are considered reproducible resources though personalities and arts, respectively, are unique. The environmental qualities cannot augment on Earth and they are non-reproducible resources except the biological resources, capturing of solar energy and expansion to cosmos.

In the last century, the global GDP has increased about four times measured by real prices, i.e. corrected for inflation. Capital has grown due to trade but also robbery under various banners, which involved high social costs. Better health and education of workers has contributed to the productivity growth but also low pay, self-employment and non-pecuniary work of women, slaves, serfs, artists and volunteers who did not benefit from the income growth. The productivity of environmental qualities refers to the extraction and use of minerals, land and other natural resources. These activities cause threats for good health and environment on the local level (e.g. diseases caused by dirty water), on the regional scale (e.g. acidification of rain) and globally (e.g. climate change and biodiversity loss). Some threats are contained (e.g. ozone layer depletion), but new ones emerge (e.g. genetic

1.1 Goals and Terms

deficiencies, lightness). Preventing degradation of environmental qualities involves continuous activities that require growing efforts. Nevertheless, welfare has grown a lot in most countries when the real income is corrected for the purchasing power and social costs. This welfare growth cannot solely be explained by the growth and allocation of capital, labour and environmental resources.

The GDP growth observed throughout the last century is attributed to the productivity growth due to technological change in the sense of development and use of new production equipment, tools and methods as elaborated by Solow and Swan in the 1950s. The technological change is driven by changes in the capital goods and the consumer goods and services aiming at private and social income often called process innovations and product innovations, respectively. Innovations are comprehended as 'doing things differently' after Schumpeter (1939, 1989: 59), which means development and new uses of technologies being equipment, constructions, tools, processes, products, services, designs, models, images and brands; the organisational and institutional changes are included if directly related to technologies. Innovations increase productivity entailing the income growth when they reduce the production costs, i.e. make it more cost-effective, and increase output, i.e. create value added. In the last decades more attention is given to knowledge as a resource of innovations. In the new growth theory, innovations would be due to the practical knowledge called know-how (reviewed for example in Helpman 2004).

Knowledge is generally comprehended in the sense of understanding events and know-how as an entrepreneurial capability of transforming knowledge into activities. Know-how is a skill. It starts with shutting laces, swimming, riding bicycle and suchlike without much knowledge. Some people catch these skills fast but others do not manage despite attempts (my mother, excellent mathematician, couldn't but I can despite my poor math). Know-how, therefore, is not solely, maybe even not primarily, about generating knowledge through understanding, research and so on because it is also about the individual skills of transforming knowledge into activities, as well as about the social capabilities to generate and combine different types of know-how for innovations. Herewith, innovations that generate welfare growth, in particular enlarge the social income and improve environmental qualities, are called sustainable innovations, synonymous to eco-innovations and the like. Underpinning the diversity of methods and conditions for sustainable innovations is the aim of this book.

Decisions about innovations are difficult to make. All innovations involve high costs and uncertain rewards at the time of decision making. In addition, innovations can undermine the vested interests labelled as 'creative destruction'. The usual reaction is opposition to such changes. The decisions about sustainable innovations cause additional difficulties. It is because environmental qualities are collective goods that are valuable to many people but difficult to distribute. In effect, all can be considered responsible but not many want to pay for these qualities when do not gain individually. It can justify inactions or gimmicks of ownership, such as quasi-privatisation of climate through compensation credits for greenhouse gasses or of biodiversity through gene banks.

Environmental qualities are also rarely reproducible though some scholars assume that the man-made capital for nature substitution is possible. The risk is these qualities are deplored when harmful technologies are used but the assessments of harms can fail when the incubation time of impacts is long. A precautionary decision making can prevent the impacts but an over-precaution can paralyse changes towards less harmful activities. There are also impact trade-offs. Improvement of one environmental quality can collide with another quality, for example, wind energy production can supply clean energy but degrade landscape. It is about the balance of positive and negative impacts, and many methods to assess pros and cons are available but all methods are only assumptions about important metrics (reviewed in Tulenheimo 1997; Frischknecht and Junbluth 2007; Ĉuĉek et al. 2012).

Innovations are driven by sense of urgency, as Frank Zappa sang 'Necessity is Mother of Inventions'. This also holds for sustainable innovations. There is sense of urgency because environmental qualities are highly demanded. The demands address the use of soil, energy and minerals, nature, space and suchlike qualities for income, safety, health well-being, culture and other private and social interests. These qualities are under pressure. For good environment, many pressures must be reduced factor four and even more in comparison with the present situation (Weizsacker et al. 1997). In particular, pressures on climate, nitrate balance and biodiversity must be reduced (Rockström et al. 2009; UNEP 2011). A widespread assumption is that the income growth and better environmental qualities are incompatible because more income would cause more impacts on the environment. This assumption is often substantiated with the 'hockey stick' graphs that present the exponentially growing impacts along with income growth throughout the last century. This assumption is challenged in this book. This is mainly based on practices because in theory income can grow along with impact reduction. For example, if sustainable innovations would increase income by 2 % a year and reduce environmental impacts by 7 % a year, the total income could increase by 50 % along with four times lower impacts in one generation.

The debate addressed in this book is about how to satisfy private and social objectives with respect to income and environmental qualities. The policymaker's perspective on this debate can be found in several books on the environmental economics. The usual viewpoint on the innovating for sustainable development is that corporations and institutions develop and use technology when invoked by demands and enforced through regulations. This perspective is also taken in my previous book *Innovations and the Environment*. This book *Theory and Practices on Innovating for Sustainable Development* adds the entrepreneurial perspectives. The proposition is that income can grow along with better environmental qualities if the sustainable innovators are encouraged. These innovators are found in the academic world, research institutes, corporations and small- and medium-scale enterprises, in technology firms as well as in 'garages and kitchens' as individual entrepreneurs, tinkerers and tacit inventors. Why and how sustainable innovations emerge in the domains of solar power, tourism, culture, sanitation, transport, consumer products, finances and renewable energy is presented.

1.2 Environmental Debate

The debate about income growth versus environment is ongoing throughout the last two centuries. It is worth a brief review because several arguments are reproduced in different terms and with nuances. An anti-cyclical sentiment in this debate can be observed: pessimists are vocal when economies grow and optimistic views on the income and environment strengthen during economic contractions. It shows that observations are multi-interpretable. The cycles refer to the productivity upswing every 40–60 years due to basic innovations – steam, rail, electricity, chemicals and computers are usually mentioned – followed by crisis, recession, revival and within each long wave a few business cycles of 15–25 years due to innovations in capital goods that undermine some vested firms and generate entrants on markets (Schumpeter (1939), 1989; Mensch 1975; Perez 2009). Inter alia, it is argued that the dotcom burst in 2001 has signalled saturation of the information and communication technologies' (ICT) long wave, which would cause the economic crisis after the financial crash in 2008 (e.g. Mecking 2008).

When steam-based industries in the early 1800s raised income but threatened rural communities, John Stuart Mill, a leading liberal economist and philosopher of that time, expressed his preference for the stationary state (Mill (1884), 1985, p. 116). In his view, the stationary state implied augmenting social and moral goods instead of using more land. Income growth would be possible and desirable but less dependent on the use of land. One could interpret it as advocacy of an immaterial growth but his term of the 'stationary state' is used as a reference to the 'steady state' meaning zero-growth or degrowth economy with distribution of wealth (Daly and Cobb 1989). The income growth due to industrialisation enabled better hygiene which reduced mortality entailing population growth. Many opinion leaders from the Mill's generation, among them Thomas Malthus, have argued that the population growth makes demands to grow, which enhances recovery from crises, but the dark side of the larger population and their demands is overexploitation of land with food shortages and mass starvation (Hodgson 2004). The view on the malicious effects of growing population and their demands is reproduced by the environmentalist opinion leaders in the Club of Rome during the chemical boom in the 1960s and in the Degrowth Declaration during the informatics boom in the 2000s (Declaration 2008). They have not remarked that the demand growth can counteract crises and that the mass starvation expected by Malthus is prevented during his lifetime through innovations on farms. These innovations enabled to generate food surplus whenever policies distributed land for farming.

Forty years later, when urbanisation and mass transport enlarged coal use, William Stanley Jevons predicted fuel scarcity because the growing fuel use would exceed the mining capacities. Based on his argumentation 150 years ago, rebound effects are pinpointed nowadays. A rebound effect means that the growing consumption of natural resource would exceed the material-saving technological change entailing resource depletion. This argument is presently often used against consumption (e.g. Alcott 2008), but the point is missed that in the nineteenth century, the UK and US coal for wood substitution has released fuel constraints, which is followed by oil for coal substitution in the twentieth century. The substitutions due to entry of new cost-effective processing technologies, so-called backstop technologies (Nordhaus 1973), are presently widely acknowledged. Backstop technologies are widespread across all businesses. They are introduced when parity with resource prices and with performance of the vested, rival technologies is attained; the rivals are considered in the sense of substituting each other, coined as "disruptive" (Christensen 2000). Such substitutions may take some time when novel technologies become gradually more cost-effective due to adaptations in production and use of goods, labelled as respectively 'learning by doing' and 'learning by using'. In general, economies responded with innovations to mineral scarcities and pollution under pressures of social demands and regulations (Rosenberg 1975). For instance, windmill electricity production on land was costly until the 2000s when it reached price parity with fossil fuel electricity production due to larger scale, less repair, better transmission and other adaptations. The coal for wood and oil for coal substitutions in the nineteenth century were not spontaneous market events, neither market arrangements of the corporates, but they were forced by authorities. The US authorities, for example, pursued successfully independence of energy imports throughout the nineteenth century. In general, economies responded to the scarcities and pollution under pressures of social demands entailing regulations (Rosenberg (1973) 1977). It resembles present debates about fuel dependency.

Fifty years later, the boom of the early 1900s inspired Pigou (1920) to elaborate on the unintended negative effects of economic activities on property rights. These effects are called the external effects. Observations of the external effects, such as sparks emitted by trains that caused fires along railroads, brought him to conclude that the rule of law is necessary for balancing of the private and social costs of activities. The rule of law could restrict railroads or impose liabilities for damage or introduce a tax to compensate the costs of victims and support restoration of their damages. The tax, it is theorised, would allocate resources efficiently. The viewpoint that social costs of hazards should be included in market prices gained support in the economic theory as internalisation of external costs. Policies on pricing of external effects through regulatory pollution charges would effectively invoke the development and use of pollution control technologies (reviewed in Nentjes and Wiersma 1987). This approach has been disseminated towards nearly all countries as environmental policy but the taxes are rarely used because of obstruction by polluters.

Modern environmentalism emerged during the chemical boom in the mid-1900s as a social movement advocating restrain in production and consumption vision with support of neo-Malthusian computer scenarios on exhaustion of natural resources and pollution (Meadows et al. 1972). The counterview is that economies can grow if impacts do not degrade the environment and non-renewable resources are substituted by renewable ones at a higher pace than consumption (Kuipers and Nentjes 1973; Solow 1973). This argumentation is presently referred to as 'decoupling' or 'dematerialisation'. It paved way during the recession of the 1980s for the global policy consensus on stewardship of the environment and economy called 'sustainable development'. Innovation is at the core of this consensus. As stated by the World Commission on Sustainable Development (WCED): 'Humanity has the ability to make development sustainable to ensure that it meets the needs of the present without compromising the ability of future generations to meet their own needs. The concept of sustainable development does imply limits – not absolute limits but limitations imposed by the present state of technology and social organization on environmental resources and by the ability of the biosphere to absorb the effects of human activities. But technology and social organization can be both managed and improved to make way for a new era of economic growth' (WCED 1987:43). Numerous innovations are envisaged, framed with lively metaphors of sustainable metabolism, tree-like business, green consumption, biomimicry and suchlike (Reijnders 1984; Winter 1987; Ayres 1989; Elkington and Burke 1990; Benyus 2002). Many innovations would be cost-effective if industries anticipate far-reaching social demands for environmental qualities (e.g. Huisingh et al. 1985; van Driel and Krozer 1988). This opinion was disputed because private and collective interests in the environment would not differ and policies would add costs without positive results, but by the 1990s, the view that policies can invoke cost-effective innovations with benefits to societies is approved by corporates and consultants (Graedel and Allenby 1995; Porter and van der Linden 1995; Hawken et al. 1999).

The ICT boom in the 1990s is accompanied with degrowth sentiments. Securing environmental qualities with radical income distribution to meet basic needs of all is propagated (e.g. McKibben 2007). This sentiment flipped again after the financial crisis in 2008. Innovations for the so-called 'green growth' are advocated, which echoes the views on innovations for sustainable development. The high-income countries associated with the Organisation for Economic Co-operation and Development (OECD) recommended subsidies for eco-innovations (OECD 2011), policy support of renewable energy is enlarged despite public debt, and the global 'green growth' policies are promoted (UNEP 2011; OECD 2013). This advocacy is underpinned with scenarios based on the tangible technologies because many are available and tested in practices. For instance, water for all human demands need not to be scarce if water catchment, recycling and desalination technologies are applied (Rosegrant 2002; Shen et al. 2008), although the water use for biodiversity requires changes in agricultural and industrial practices (Alkemade et al. 2009). Economies would not run out of minerals if the available waste prevention and recycling technologies are used (Allwood et al. 2011) though availability of some minerals remains critical because of overuse, such as the use of phosphate (Cordell 2010). The agricultural pressures on biodiversity, water for health and food production and climate change would be managed if the available technologies are implemented (Hajer 2012), and the global energy use could be free of fossil fuels by 2050 if energy-efficient and renewable energy technologies are applied (WWF et al. 2011). Sustainability notion would be the driver of innovations in businesses (Nidumolu et al. 2009), and clean, green growth would counteract the economic crisis and improve environmental qualities if enforced through 'smart' regulations, which means using taxes on polluting activities and technology forcing policies (Jänicke 2011; Swilling 2013). The financial crisis would provide opportunity for welfare growth because it enables to shift resources towards sustainable innovations (Perez 2013). Smart regulation for clean development is an appealing vision. These arguments well known from the past debates on sustainable development are repeated having experiences with cleaner technologies.

A recurring argument in this debate about economy and environment is that innovations should be steered towards welfare and environment but how to steer is disputed. Market would steer spontaneously in the right direction if social costs of environmental degradation are internalised in prices, policies would generate social arrangements that lead transition to sustainability, more knowledge would resolve problems because generate and disseminate solutions, stakeholders' involvement and citizens' participation would invoke proper decisions, and so on. It is tempting to envision the roadmap for sustainable development. With reference to the Lao Tzu proverb 'those who have knowledge don't predict and those who predict don't have knowledge', predicting panacea usually proved to be wrong or disastrous. Creating opportunities for the diversity of sustainable innovators could be a more fruitful approach. Incidence is useful when prophets are imperfect.

1.3 Innovating for Environment

The conventional economist viewpoint on innovations is that, given competition, private and public demands invoke technology development if innovation rents are expected; the rent is income from capital minus costs. The rents are attained when innovators sell profitably and users gain due to lower costs or higher value of the innovations compared to available technologies. The rents are uncertain because knowledge can be deficient, customers' demands can change, competitors can capture a market share, and others (Stoneman 1983). The innovation process would evolve in a planned manner in phases: research delivers a novelty (invention) and development proofs that the novelty works (designs), demonstration shows match with demands (model), small-scale production can start (pilot), and sales be launched (market introduction). Herewith, costs are made during several years without sufficient income to cover them. The innovation rents can be generated during the dissemination of innovations (diffusion) but the risk is that these costs are not covered. This presentation of the innovation process called demand-pull model of innovation has disseminated across management handbooks, standard works on policies and in statistics and so on. It became the mainstream viewpoint on the innovation sources, although it is underpinned that innovations are rarely made on demand because the demand factors change and they are uncertain and that innovations mainly depend on the entrepreneurial capabilities (Mowery and Rosenberg 1979).

The demand-pull model seems attractive because it suggests that policymakers and managers can steer the innovations processes towards priorities set well ahead of the realisation. Various steering strategies are assumed to be successful. The large-scale, isolated research centres that generate basic knowledge are considered effective in the 1960s because industries would pick and use the basic knowledge for applications. A decade later it is argued that the German technology forcing regulations have invoked the industrial research and development which generated innovations in this country, but the Japanese innovations would be due to the intensive business–government cooperation and supply chain management. In the 1990s, the protectionist, autoritarian policies are considered successful for the 'Asian Tigers' (Singapore, Taiwan and South Korea), as well as the liberalisation policy and market-based corporate management in the former Communist countries and India. In the 2000s, the mission-oriented, specific public funding and policy leadership in the United States are pinpointed as driver of innovation, as well as the generic excellence in education and incentives for creativity in the Scandinavian countries. If all these arguments are valid, policy and management of innovations are time and culture dependent but the validity is also disputable because biased by scholarly focusses and political preferences. These argumentations give food for thoughts about options and effects of innovation policies and management but prophecies about successes are contentious and handbooks on how to do it seem to be 'lecturing birds on how to fly' as Taleb has remarked (Taleb 2012, Chap. 3).

The demand-pull model could be reflections on innovation processes in the large industries and research centres that are accessible for research. The model could also be valid when knowledge is scarce but could have limited validity when knowledge becomes widely available because millions of people study and research. Meanwhile, it is observed that the share of small- and medium-size enterprises in all innovations has increased during the last 50 years; that public institutions, universities, services and individuals also innovate; and that the large corporations possess many patents for protecting their core business rather than for innovations and they buy the innovating firms rather that innovate. There are many innovation sources.

In the emerging knowledge-based societies, innovations are often derived from the collective or open knowledge sources and from private expert opinions. Innovators combine this knowledge, which is available within various networks of interests, with concepts (designs) that are tested under various conditions. Herewith, innovating is about the entrepreneurial skills of sourcing and using knowledge for experimentation onto marketable products. This know-how enlarges through learning by doing and learning by using, and numerous successful innovators have rather limited formal knowledge but excellent know-how skills and persistence in the experimentation (Petrovsky 1996), among them the founders of Apple and Facebook. A model of the innovator behaviour suggests that the innovators could embrace two different archetypes of skills: ones who envision and experiment and ones who measure and engineer. The former, 'Stochasts', use incidence, the latter, 'Cartesians', use metrics. The Stochasts who invent tacitly would be generally more successful innovators and the Cartesians who typically emulate more successful during diffusion (Allen 1988). This view on innovations assumes that the prime innovators' skill is recognizing incidental opportunities and creating a nexus of interests, an innovation network. Such network model of innovations, however, is difficult to verify. It is not intended to define and specify the network model of innovations in this book but underpin the diversity of successful sustainable innovators. These can be individuals with outstanding skills, excellent experts in small- and medium-size enterprises, dedicated volunteers in social organisations and professionals in policymaking, artists, designers and project developers striving for new concepts, as well as scientists and engineers in research and development centres and in corporations.

The hypothesis about the innovation networks suggests that the personal skills of entrepreneurs are essential because they enable to take advantage of market opportunities that emerge due to diversity of events. It is often called creativity. The generation of know-how for innovating is also a fuzzy process of social interactions which involves innumerable knowledge-related exchanges between various interests, called knowledge spillover. These are needed to gain expertise, test quality, find customers, get funding and involve labour and other entrepreneurial activities. The interactions are often presented as an innovation system. The term system suggests planning but valuable interactions during work and leisure are generally incidental and the value of knowledge spillovers stochastic (Freeman 1996). Many inventions are serendipitous, which means incidental discoveries. Conditions for the valuable interactions can be defined as being the abundance and accessibility of knowledge, diversity of people involved, freedom of expressions and creativity and so on. The interactions imply that innovating is a social process based on know-how. This quasi-collective character of innovation processes enables to reduce costs and risks but it also implies that the proceeds cannot be claimed unless an innovator is entitled for the monopoly use of inventions. Such entitlements are given by an authority as patents and copyrights. The entitlements attract investors who seek innovation rents because they can claim ownership but these entitlements also impede knowledge spillovers, and therefore, the innovations spur. Many scholars argue to restrict or abolish these entitlements. The restrictions are needed when the entitlements impede valuable innovations or cause social and environmental harm but the abolishment can impede investments in innovations because private gains are missed. Alternatives are innovations in the public domain and open source innovations, but public institutions are often risk avoiding and the open source innovations are scarce, as yet. How to satisfy the investors, innovators and societal interests is an unresolved dilemma.

Innovating with the aim to foster environmental qualities adds to this complexity because not solely the innovation process is quasi-collective but also many environmental qualities are quasi-collective goods. Rewards are uncertain because people rarely pay for the benefit of collective interests without direct private gains unless they generate social demands for such improvements (Jaffe et al. 2005; Krozer 2008). The social demands for innovations are expressed via market, through policy and behavioural codes, but the expressions are imperfect. A lot is studied on policies to improve these expressions. The neoclassic and evolutionary theories are reviewed in Jaffe et al. (2005) and Ruttan (2012); Krozer and Nentjes (2006) have also included the behavioural viewpoint. Beneath, only a few relevant notions are highlighted and confronted with some experiences. The assumption in the mainstream neoclassic approach is that the demands for good environment drive up the resource and pollution prices, which enables to bear the costs of innovation because rewards can generate higher rents. Other views confirm that the costs and proceeds are important factors but underline that other factors than the prices drive innovators to act. The basic assumptions in all theories are that all private interests are knowledgeable and that authorities enforce without risk of cheating or robbing, verily the place to be.

The neoclassic theory is focused on the price mechanism for allocation of scarce resources and incentive for innovations. Two viewpoints on how to reach the optimal price of environmental qualities can be found. One viewpoint is that negotiations between private interests involved in a collective good – nowadays called stakeholders – attain the optimal price independent of the initial distribution of that good if the stakeholders decide freely and if their transactions are costless and timeless (Coase 1963). This argumentation is the basis for the policy of 'self-regulation' through gentlemen's agreements between stakeholders (covenants). These are introduced in several countries, international negotiations and business supply chains; the Netherlands has championed gentlemen's agreements with hundreds of covenants and idle regulations during the periods of agreements. The experiences are that covenants generally fail to fulfil agreements and are used to postpone policies (Ashford 1996). Another opinion within the neoclassic theory is that authority is needed because an imperfect distribution of property rights causes deficient prices. The authority would make the users of common good pay for it, so-called polluter pays principle. The usual instrument is pollution permit but economists generally recommend a charge that reflects the marginal social costs and benefits of pollution (Baumol and Oates 1975). The charges would invoke more innovations because permits do not require actions beyond the permit limit, whereas the charges trigger additional pollution reduction, but the permits are widely used and the charges are scarce.

A different starting point is that innovations are results of cultural constructs in science and engineering. These constructs expressed as engineering scripts and designs generate patterns in technology development. In economics, this viewpoint is reflected in the evolutionary theory. This is focused on knowledge. Innovations would be generated due to the innovators' search for options with useful engineering characteristics followed by selection of the options that match the innovator's aims (Dosi and Orsenigo 1988). When innovations disseminate they generate increasing returns to scale, which increase productivity across businesses, called spin-off. Those innovations that disseminate at a high rate entailing a large market share dominate because interests must link to them. A technological system is vested. It also means that the deficient technologies can cause the lock-in deficiencies. A vested lock-in deficiency is difficult to change even if superior alternatives are available because many interests are bound to it. Authorities are needed to resolve such lock-in (David 1975; Arthur 1989). The authorities, therefore, should define targets to be achieved and lead towards the path-breaking innovations through funding and management (Mazzucato 2011). Environmental policies would cause lock-in when they aim to reduce pollution in the short run using end-of-pipe technologies instead of preventing pollution through process-integrated technologies that are cost-effective in long run. Resolving this deficient lock-in could be through technology forcing; it is enforcing technologies that are not available at the moment of policymaking (Kemp 1995), but this policy is rarely used because it risks failure if sound technologies remain unavailable. It would also need an array of policy interventions (Arentsen et al. 2001).

This viewpoint became popular in policies labelled as transition management. Strict policy demands, herewith, are considered contra-productive because they invoke deficient solutions. Instead, broad societal changes are pursued aiming at optimal solutions in the long term. These are so-called 'win-win'. Firms would seek the 'win-win' if they are supported in the search and selection of innovations instead of regulations (e.g. Willard 2002; Esty and Winston 2006; Savitz 2006). Managers would lead transitions if they have the freedom to choose solutions instead of complying with regulations (Holliday et al. 2002), businesses would spontaneously generate technologies that provide highly valued products (McDonough and Braungart 2002), know-how on pollution prevention would create new successful businesses (Bradfield and Nogrady 2010), and so on. All these are rare events. To the contrary, the advocacy of the optimal solutions in the long run instead of social demands for innovations in the short run is employed to temper innovations if perceived contrary to the vested interests even though many innovations are socially net beneficial, e.g. in renewable energy.

The third viewpoint is focused on behaviour of people and organisations. In the behavioural view, it is assumed that organisations operate within the vested patterns and hierarchies in which decisions that are perceived important are controlled high in the hierarchies and the less important decisions are delegated to the lower levels (Cyert and March (1963), 1988). Policymakers, herewith, are usually bound to internal procedures that are vested and evolve largely independent of the socio-economic development perceived external to the policy organisation. Policies would generally impede innovations because the organisational changes in institutions evolve slower than the technological changes (Cohen et al. 1972). Within the organisations, it is argued, individuals rarely act based on the elaborated assessments of pros and cons but decide fast and intuitively. It is called intuitive when derived from habits and in conformity with the socially accepted norms. Hence, decision making is biased by the vested habits and interests. The behaviour is typically non-innovative, geared to the vested interests and therefore, risk avoiding (Kahneman (2002) 2011). Game theoretical views and behavioural experiments, however, suggest that well-designed social mechanism disseminate and invoke behavioural changes with respect to the private and social interests albeit conditions for the dissemination are disputable (Thaler and Sustein 2008).

The argument with regard to an environmental quality is that it is overused when the private interests dominate because the private gains of using such quality at a cost of society are beneficial. This free-riding behaviour persists if uncorrected by an authority (Hardin 1968). Another argumentation is that collective arrangements based on the groups' interests prevent this overuse. This option ushers moral codes for group, which restricts the free riding and regulates individual behaviour in use of the common good. These arrangements would foster environmental qualities due to the ethical behaviour with respect to the collective interests (Ostrom 1998). A drawback would be that the introduction of such arrangements involves laborious transactions (Barbier 2011).

The presented views collide when confronted with a tangible environmental issue. For example, having to decide about pollution reduction of a water body, one neoclassic advisor would insist on the property rights with the non-interventionist policy aiming to give freedom of negotiations between stakeholders, whereas another neoclassic advisor would favour a policy that puts a pollution price aiming to internalise the external effects. Doing both can cause contradictions. The evolutionary experts would insist on cooperation between stakeholders in pursuing knowledge about alternatives and selection of solutions that are optimal for all in the long run but this process can go on for long without results, which reinforces the vested interests. Whether the ultimately selected solutions only appear to be optimal but fail after the decision making and how to avoid the biased decision making caused by power imbalances during negotiations are unsolved issues. The behavioural consultants would not be optimistic about any policies in support of innovations because institutional behaviour is perceived risk avoiding. They would argue that those policies create conditions for sustainable innovations as well as protect the vested interests that oppose such changes. These consultants would recommend support of the victims of pollutions because the victims have interests in innovations for the cleaner water body but what instruments can be used and whether these generate social benefits are unclear.

More policy or markets are not necessarily the solutions. Less policy and managerial interventions but far-reaching individual and social demands for environmental qualities with freedom of actions could also do. There is reason for optimism about the private and social interests for sustainable innovations if policies foster markets of sustainable innovations, secure fair play competition and put prices on environmental degradation that reflect social costs. First, firms become capable in addressing environmental issues. This capability is reflected in the corporate social responsibilities management. This is management high in the corporate hierarchy that can push innovative ideas through administration with funding and expertise aside notable examples of mismanagement through greenwashing. It is in line with the advocacy of 'foolishness' in organisations (March 1971). Second, costs can be reduced and environmental qualities improved through prudent resource use, preventing environmental impacts, turning waste into products, recovery of degraded spaces and so on. The collective interests for good environment can be linked to private interests for value-adding services. Third, businesses can emerge due to demands for products and services with credible qualities. Although there is trade-off between prices and qualities, such demands enable companies to innovate using ethical attributes in products. Fourth, firms, social organisations and policymakers generate local networks with potential to communicate globally through modern media, referred to as the 'distributed economy' (Johansson et al. 2005). This enables the use of regional qualities for entrepreneurial activities. Fifth, people learn individually to develop and make products and services instead of solely consuming them, which makes them producers and consumers in one, expressed by Toffler (1980) as prosumers. More encouraging possibilities can be found. Business models also diversify. The conventional model of the firms with capital participation is challenged by networks of professionals, service sharing, social ventures, cooperatives, consumers' participation, trusts and so on. Property rights have also been redefined through open source innovations, crowdfunding, crowdsourcing and other methods. There are many fascinating initiatives from the innovators' perspective.

1.4 Content of the Book

The diversity of innovators pursuing better environmental qualities is the thread in this book. The first two chapters underpin with statistical data that the innovation-based income growth improves environmental qualities if policies foster innovations rather than protecting the vested interests. These two chapters provide the context for nine cases that illustrate sustainable innovators. The user innovators in solar power, tacit inventors in tourism, artists that combine nature and culture, and technology suppliers in sanitation illustrate that the small-scale innovators dominate in the present knowledge-based societies. Cases of project developers in offices, corporations for ethical consumption, financing sustainable innovations, and local energy initiatives show that beneficial system changes are feasible in the short run if policies do not support rent-seekers. In comparison with much larger subsidies in the United Stated, the European feed-in tariffs for renewable energy mitigate climate change more cost-effectively. The cases are presented in an ascending order of complexity: from single products, through systems and markets, to businesses and policies. The book ends with conclusions from the policymaker's perspective.

Are income growth and environmental impacts compatible? The book starts in Chap. 2 with the debate about income growth versus environmental impacts. It is underpinned with statistical data that the quasi-autonomous technological change reduces environmental impacts with income growth in the high-income countries.

How are sustainable innovations induced? Markets of sustainable innovations and impediments are shown in Chap. 3. The global demand and impediments for sustainable innovations are assessed. The demands for environmental qualities induce sustainable innovations in the areas of material resource, pollution control, ethical consumption, ecosystem services and cultural attributes, but policy support of the vested, rival interests is as large as the markets.

Can consumers innovate in complex products? The case in Chap. 4 is about the user innovations in solar-powered boats. It is shown that consumers, being volunteers, technician and students, develop advanced technologies and generate high-tech markets.

Can economic peripheries use tacit knowledge for innovations? This issue is elaborated in Chap. 5. It is about tacit know-how of local social organisations, in small- and medium-size enterprises and in institutions used for sustainable innovations in tourism. The regional network and cluster policies are compared.

Can artists foster valuable but presently unrecognised environmental qualities? This is discussed in Chap. 6. Possibilities of income generation through arts services and using arts for imagining valuable innovative uses of environmental qualities are addressed.

Can lock-in into a costly technology be resolved? Technology supplies for four sanitation systems are shown in Chap. 7. The dominant system is effective but costly. Alternatives can be cost-effective but have other drawbacks. Tuning to the community interests can be fostered.

Can project developers reduce traffic congestions? Chapter 8 compares the life cycle costs of four alternative office systems. The system embraces office work with commuting. The social benefits of the office alternatives are underpinned.

Do consumers induce ethical consumption or companies seeking credibility in sales? This question is answered in Chap. 9 based on experiences with ethical attributes in the consumers' purchases and with the trendsetting large corporation.

Do public funders and private investors foster sustainable innovations? This issue is addressed in Chap. 10 based on market data and opinions of sustainable investors and innovators about each other and about policies.

What are business opportunities and impediments for innovators in energy services? Chapter 11 underpins that many local initiatives capture opportunities on energy markets when missed by vested interests despite policies that protect the vested interests.

What are the driving factors in renewable energy? This is discussed in Chap. 12 based on statistical data for the European Union and comparison of the policy instruments in the European Union and the United States and regional experience.

Chapter 13 ends with conclusions about sustainable innovations from the policy-makers' perspective.

The chapters can be read separately without mathematical background because the calculations are explained and the results are presented (there are only a few formulas). The relevant costs and benefits are presented in USD and euro. For convenience, for one euro one point three US dollars is used (€ 1 = USD 1.3). If the real values are calculated, the annual average conversion rates are used.

References

Alcott, B. (2008). Historical overview of the Jevons paradox in literature. In J. M. Polimeni, K. Mayumi, M. Giampietro, & B. Alcott (Eds.), *The Jevons paradox and the myth of resource efficiency improvements* (1st ed., pp. 7–78). London: Earthscan.

Alkemade, R., van Oorschot, M., Miles, L., Nellemann, C., Bakkenes, M., & ten Brink, B. (2009). GLOBIO3: A framework to investigate options for reducing global terrestrial biodiversity loss. *Ecosystems, 12*, 374–390.

Allen, P. M. (1988). Evolution, innovation and economics. In G. Dosi, C. Freeman, R. Nelson, G. Silverberg, & L. Soete (Eds.), *Technical change and economic theory* (1st ed., pp. 95–119). London/New York: Pinter Publishers.

Allwood, J., Ashby, M. F., Gutowski, T. G., & Worrell, E. (2011). Material efficiency: A white paper. *Resources, Conservation and Recycling, 55*, 362–381.

Arentsen, M. J., Dinica, V., & Marquart, E. (2001). Innovating innovation policy. Rethinking green innovation policy in evolutionary perspective. *Économies et Sociétés, Série Dynamique technologique et organisation, 4*(4), 563–583.

Arthur, W. B. (1989). Competing technologies, increasing returns and lock–in by historical events. *The Economic Journal, 99*, 116–131.

Ashford, N. A. (1996). The influence of information–Based initiatives and negotiated environmental agreements on technological changes. In C. Carraro & F. Lévêque (Eds.), *Voluntary approaches in environmental policy* (1st ed., pp. 137–150). Dordrecht: Kluwer Academic Publishers.

Ayres, R. U. (1989). Industrial metabolism. In J. H. Ausubel & H. E. Sladovich (Eds.), *Technology and environment* (1st ed., pp. 23–49). Washington, DC: National Academy Press.

Barbier, E. B. (2011). Transaction costs and the transition to environmentally sustainable development. *Environmental Innovation and Societal Transitions, 1*, 58–69.

Baumol, W. J., & Oates, W. F. (1975). *The theory of environmental policy* (1st ed., pp. 14–55). Englewood Cliffs: Prentice–Hall.

Benyus, J. M. (2002). *Biomimicry* (1st ed.). New York: HarperCollins.

Bradfield, M. J., & Nogrady, B. (2010). *The sixth wave* (1st ed.). Sidney: Random House.

Christensen, C. M. (2000). *The innovator's dilemma* (1st ed.). New York: Harper Business.

Coase, R. ((1963) 1972). The problem of Social Cost. Reprint in R. Dorfman & N. S. Dorfman (Eds.), *Economics of the environment: Selected readings* (1st ed., pp. 142–171). New York: W.W. Norton & Company.

Cohen, M. D., March, J. G., & Olsen, J. G. (1972). A garbage can model of organizational choice. *Administrative Science Quarterly, 17*(1), 1–25.

Cordell, D. (2010). *The story of phosphorus* (Linköping studies in arts and science no. 509). PhD thesis, Department of Water and Environmental Studies Linköping University, Linköping, pp. 122–128.

Ĉuĉek, L., Klemeś, J. J., & Kravanja, Z. (2012). A review of footprint analysis tool for monitoring impacts on sustainability. *Journal of Cleaner Production, 34*, 9–20.

Cyert, R. M., & March J. G. ((1963) 1988). A behavioural theory of organisational objectives. In R. M. Cyert (Ed.), *The economic theory of organisation and the firm* (pp. 125–150). New York: New York University Press.

Daly, H. E., & Cobb, J. B. (1989). *For the common good* (1st ed.). Boston: Beacon.

David, P. A. (1975). *Technical choice innovation and economic growth* (1st ed.). New York: Cambridge University Press.

Declaration (2008, April 18–19). Declaration of the Paris 2008, Conference, *Economic de-growth for ecological sustainability and social equity conference*, Paris.

Dosi, G., & Orsenigo, L. (1988). Coordination and transformation: An overview of structure, behaviours and change in evolutionary environments'. In G. Dosi, C. Freeman, R. Nelson, G. Silverberg, & L. Soete (Eds.), *Technical change and economic theory* (1st ed., pp. 13–37). London/New York: Pinter Publishers.

Ecofys, W. W. F., & OMA. (2011). *The energy report: 100% renewable energy by 2050*. Gland: WWF.

Elkington, J., & Burke, T. (1990). *Groen zaken doen* (1st ed.). Amsterdam: Maarten Muntinga.

Esty, D. C., & Winston, A. S. (2006). *Green to gold* (1st ed.). New Haven: Yale University Press.

Ferguson, N. (2008). *The ascent of money* (1st ed.). London: Penguin.

Freeman, C. (1996). The greening of technology and models of innovation. *Technological Forecasting and Social Change, 53*, 27–39.

Frischknecht, R., & Junbluth, N. (2007). *Implementation of life cycle assessment methods*. Dubendorff: EMPA.

Graedel, T. E., & Allenby, B. R. (1995). *Industrial ecology* (1st ed.). Englewood Cliffs: Prentice Hall Publishers.

Hajer, M. (2012). *Roads from Rio20+, pathways to achieve global sustainability by 2050*. Bilthoven: PBL, Netherlands Environmental Assessment Agency, mimeo.

Hardin, G. (1968). The tragedy of the commons. *Science, 162*(3859), 1243–1248.

Hawken, P., Lovins, A., & Hunter Lovins, L. (1999). *Natural capitalism* (1st ed.). New York: Little Brown and Company.

Helpman, E. (2004). *The mystery of economic growth* (1st ed.). New York: The Belknap press of the Harvard University Press.

Hodgson, G. M. (2004). Malthus, Thomas Robert (1766–1834). In D. Rutherford (Ed.), *Biographical dictionary of British economists* (pp. 1–6). Bristol: Thoemmes Continuum. http://www.geoffrey-hodgson.info/user/image/bdbe-malthus.pdf. Visited 30 Apr 2013.

Holliday, C. O., Schmidheiny, S., & Watts, P. (2002). *Walking the talk* (1st ed.). Sheffield: Greenleaf Publishers.

Huisingh, D., Martin, L., Higler, H., & Seldman, N. (1985). *Proven profits from pollution prevention*. Washington, DC: Institute for Local Self-reliance, North Caroline, State University, mimeo.

References

Jackson, T. (2011). *Prosperity without growth* (1st ed.). London: Routledge.
Jaffe, A. B., Newell, R. G., & Stavins, R. N. (2005). A tale of two market failures: Technology and environmental policy. *Ecological Economics, 54*, 164–174.
Jänicke, M. (2011). *"Green Growth": From a growing eco-industry to a sustainable economy*. Berlin: Freie Universitat Berlin, Environmental Policy Research Centre, mimeo.
Johansson, A., Kisch, P., & Mirata, M. (2005). Distributed economy – A new engine of innovations. *Journal of Cleaner Production, 13*, 971–979.
Kahneman, D. ((2002), 2011). *Thinking, fast and slow* (1st ed.). London: Penguin.
Kemp, R. (1995). *Environmental policy and technical change*. Ph.D. thesis, MERIT, Universiteit van Maastricht.
Krozer, Y. (2008). *Innovations and the environment* (1st ed.). London: Springer press.
Krozer, Y., & Nentjes, A. (2006). An essay on innovations for sustainable development. *Environmental Science, 3*(3), 163–174.
Kuipers, S. K., & Nentjes, A. (1973). Pollution in a neo-classical world: The classics rehabilitated? *The Economist, 121*(1), 52–67.
March, J. G. ((1971) 1989). The technology of foolishness. In J. G. March (Ed.), *Decision and organisation* (1st ed., pp. 253–265). Oxford: Basil Blackwell.
Mazzucato, M. (2011). *The entrepreneurial state* (1st ed.). London: Demos.
McDonough, W., & Braungart, M. (2002). *Cradle to cradle* (1st ed.). New York: North Point Press.
McKibben, B. (2007). *Deep economy* (1st ed.). New York: Henry Holt and Company.
Meadows, D. H., Meadows, D. L., Randers, J., & Behrens, W. W., III. (1972). *The limits to growth* (2nd ed.). New York: Universe Books.
Mecking, E. (2008). *Deflatie in aantocht* (1st ed.). Amsterdam: Mets & Mets.
Mensch, G. (1975). *Stalemate in technology* (1st ed.). Cambridge, MA: Ballinger Publishing Company.
Mill, J. S. ((1884) 1985). *The principles of political economy* (Book IV, Chapter VI). New York: Penguin Classics.
Mowery, D., & Rosenberg, N. (1979). The influence of market demand upon innovations; a critical review of some recent studies. *Research Policy, 8*, 102–153.
Nentjes, A., & Wiersma, D. (1987). Innovation and pollution control. *International Journal of Social Economics, 15*(3 – 4), 51–70.
Nidumolu, R., Prahalad, C. K., & Rangaswami, M. R. (2009). Why sustainability is now the key driver of innovation. *Harvard Business Review*, September, 3–10.
Nordhaus, W. (1973). The allocation of energy resources. *Brookings Papers on Economic Activity*, 3:1973. http://www.brookings.edu/~/media/projects/bpea/1973%203/1973c_bpea_nordhaus_houthakker_solow.pdf. Visited 20 Nov 2014.
OECD. (2011). *Toward green growth*. OECD Ministerial Council Meeting on 25–26 May 2011, Paris.
OECD. (2013). *Green growth indicators*. Download 22 Sept 2013, Paris.
Ostrom, E. (1998). A behavioural approach to the rational choice theory of collective action. *American Political Science Review, 92*(1), 1–22.
Perez, C. (2009). *Technological revolutions and techno-economic paradigms*. Tallin University, Tallin, mimeo.
Perez, C. (2013). Unleashing a golden age after the financial collapse. Drawing lessons from history. *Environmental Innovations and Societal Transition, 6*, 9–23.
Petrovsky, H. (1996). *Invention by design* (1st ed.). Cambridge: Harvard University Press.
Pigou, A. C. (1920). *The economics of welfare* (1st ed., pp. 115–117). London: Macmillan.
Piketty, T. (2013). *Capital in the twenty-first century* (1st ed.). Cambridge, MA: The Belknap Press of the Harvard University.
Porter, M. E., & van der Linden, C. (1995). Green and competitive: Ending the stalemate. *Harvard Business Review*, September/October, 119–134.
Reijnders, L. (1984). *Pleidooi voor een duurzame relatie met het milieu* (1st ed.). Amsterdam: Van Gennep.

Rockström, J., Steffen, W., Noon, K., Persson, A., Chapin, F. S., III, Lambin, E., Lenton, T. M., Scheffer, M., Folke, C., Schellnhuber, H., Nykvist, B., de Wit, C. A., Hughes, T., van der Leeuw, S., Rodhe, H., Sörlin, S., Snyder, P. K., Costanza, R., Svedin, U., Falkenmark, M., Karlberg, L., Corell, R. W., Fabry, V. J., Hansen, J., Walker, B. D., Richardson, K., Crutzen, P., & Foley, J. (2009). Planetary boundaries: Exploring the safe operating space for humanity. *Ecology and Society, 14*(2):32.

Rosegrant, M. W. (2002). Global demand and supply projection. Part 2 Results and prospects to 2025. *Water International, 27*(2), 170–182.

Rosenberg, N. ((1973) 1977). Innovative responses to material shortages. *American Economic Review, 63*(2), 11–18. Reprint in R. Dorfman, & N. S. Dorfman (Eds.), *Economics of the Environment* (1st ed., pp. 390–399). New York: W.W. Norton & Company Inc.

Rosenberg, N. (1975). Technological innovation and natural resources: The niggardliness of nature reconsidered. In N. Rosenberg (Ed.), *Perspectives on technology* (pp. 229–259). Cambridge, MA: Cambridge University Press.

Ruttan, V. W. (2012). Source of technical change: Induced innovation, evolutionary theory, and path dependency. In A. Grübler, N. Nakicenovic, & W. D. Nordhaus (Eds.), *Technological change and the environment, resources for the future Washington DC* (pp. 9–39). Laxenburg: International Institute for Applied Systems.

Savitz, A. W. (2006). *The triple bottom line* (1st ed.). San Francisco: Jossey-Bass.

Schumpeter, J. A. ((1939) 1989). *Business cycles* (4th ed.). Philadelphia: Porcupine Press.

Sen, A. (2009). *The idea of justice* (1st ed.). London: Penguin.

Shen, Y., Oki, T., Utsumi, N., Kanae, S., Hanasaki, N. (2008). Projection of future world water resources under SRES scenarios: Water withdrawal. *Hydrological Sciences Journal, 53*(1), 11–33. http://dx.doi.org/10.1623/hysj.53.1.11

Solow, R. M. ((1973) 1977). The economics of resources or the resources of economics. Reprint in R. Dorfman, & N. S. Dorfman (Eds.), *Economics of the environment* (1st ed., pp. 354–370). New York: W.W. Norton & Company Inc.

Stiglitz, J. E., Sen, A., Fitoussi, J. -P. (2010). *Report by the Commission on the Measurement of Economic Performance and Social Progress*. http://www.stiglitz-sen-fitoussi.fr/documents/rapportanglais.pdf. 2 Dec 2013.

Stoneman, P. (1983). *The economic analysis of technological change* (1st ed.). Oxford: Oxford University Press.

Swilling, M. (2013). Economic crisis, long waves and the sustainability transition: An African perspective. *Environmental Innovation and Societal Transitions, 6*, 96–115.

Taleb, N. N. (2012). *Antifragile* (1st ed.). New York: Random House.

Thaler, R. H., & Sustein, C. R. (2008). *Nudge, improving decisions about health, wealth and happiness* (1st ed.). New Heaven/London: Yale University Press.

Toffler, A. (1980). *The third wave* (1st ed.). New York: Bantam Books.

Tulenheimo, V. (1997). *A comparative study of total cost assessment methods*. Helsinki: VTT. mimeo.

UNEP. (2011). *Green economy report*. New York: UNEP.

van Driel, P., & Krozer, Y. (1988). Innovation and preventive environmental policy. In F. Dietz (Ed.), *Environmental policy in a market economy* (1st ed., pp. 78–92). Wageningen: Pudoc.

von Weizsacker, E., Lovins, A. B., & Hunter Lovins, L. (1997). *Factor four: Doubling wealth, halving resource use* (1st ed.). London: Earthscan.

WCED, World Commission on Environment and Development. (1987). *Our common future* (1st ed.). Oxford: Oxford University Press.

Willard, B. (2002). *The sustainability advantage* (1st ed.). Gabriola Island: New Society Publishers.

Winter, G. (1987). *Das Umweltbewusste Unternehmen* (1st ed.). München: BAUM AG.

Chapter 2
Income Growth and Environmental Impacts

Are income growth and environmental impacts compatible? It is underpinned how innovations generate income and reduce environmental impacts given demands for environmental qualities. The discussion starts with the IPAT model. Empirical tests of the model indicate a decoupling of environmental impacts and income growth throughout the last century. In nearly all high-income countries, the national income and private consumption have grown along with slower increase of environmental impacts and in several countries with a decrease of the impacts. This cannot fully be explained by higher international resource prices, outsourcing of the high-impact activities to the low-income countries and countries' policies. Production structure changes fast towards services, but the share of services in consumption increases slowly. The observed relative and absolute decoupling are explained by innovations, which involve circulation of innovation rents. The resource-reducing, cost-saving process innovations reduce environmental impacts and generate innovation rents that are allocated in labour for the value-adding product innovations entailing income growth. The innovation rents approximate 13 % of the global GDP, which is sufficiently large to explain this process. The innovation-based economic growth generates better environmental qualities, given the population.

2.1 IPAT Model

After introducing the conventional environmentalist argumentation about impacts of the income growth and testing it with empirical data, a theory on the autonomous impact-reducing technological change is presented. The technologies are considered methods of production and consumption, disembodied as know-how and embodied in equipment, construction, tools, products and so on. In line with the mainstream economic view, innovations are pursued if innovation rents are expected, which is if the expected sales exceed costs, and the sales are expected if the users expect value addition. This analysis is focused on the labour and material with associated pollution. The capital resources are largely neglected, which is hardly an omission in this discussion except for the valuables as gold and suchlike.

The environmentalist viewpoint on the causes of environmental impacts is compounded in the IPAT model (reviewed in Chertow 2001). The starting point is that population causes environmental impacts denoted in the model as I. The population is defined as number of people with a fixed consumption pattern denoted as P. The growing population causes more impacts because each newborn person consumes an amount of environmental qualities throughout lifetime. The environmental impact is function of population when, k, is conversion of consumption per person into the impact (Ehrlich and Holdren 1971):

$$I = k \cdot P \quad (2.1)$$

For example, P people use X ton of oil with k conversion for ton oil per person to Y ton CO_2 and SO_2, having Z greenhouse respectively acidification impacts.

This model is refined with regard to the income and consumption patterns. Two more variables are introduced. One is consumption given income because people consume products and services with different environmental impacts. A typical consumption described as affluence, A, embraces all goods used during lifetime of a person. To meet this, capital, labour and environmental qualities are used as resources for activities throughout the life cycles of goods embracing extraction, production, distribution, consumption and disposal. When incomes grow, consumption volume increases and consumption pattern may change. The second variable is technology. Given the population and consumption per person, the resource use in the life cycles of goods changes, which causes impacts to decrease or enlarge described as technology, T. Impacts are due to number of people, consumption per person in monetary terms and technology performance per consumption unit in physical terms, T. It is formally

$$I = P \cdot A \cdot T \quad (2.2)$$

The simplicity of the IPAT model is attractive. It is simple as long as each variable is considered as being independent of another, but if assumed that all variables interact (Alcott 2010), the model becomes complex, possibly unsolvable.

The simple IPAT model can be elaborated to specify what variables influence changes in consumption and technologies. Therefore, the IPAT model is reformulated into four variables:

$$I = P \cdot \frac{a}{p} \cdot \frac{m}{a} \cdot \frac{i}{m} \quad (2.3)$$

The population, P, can grow or shrink by number of persons. The affluence per person, a/p, is expressed as an increase or decrease of income, production output or expenditures per person. It is measured in the monetary terms. Given the material properties, the volume of material use per income, m/a, indicates the effects of innovations on materials. The material use is measured in physical terms. This variable is an aggregate of several types of innovations. Given the capital, three types of innovations are important for further analysis. One type is the labour for materials substitution entailing a high share of labour-intensive activities in production and consumption. These are due to innovations for services. Given the income, the material use decreases. Note that it is not about the labour for nature substitution, which is a rather odd concept because the properties of nature are generally unique and poorly recoverable when degraded. Another type of innovations covers more effective use of materials as (input) needed to realize income (output). If the material use decreases per unit income, the impacts decrease. The third type of innovations is the labour addition to the materials. This type of innovations adds value entailing more income, which is called that materials are used more efficiently. The value added per material unit can be very high, for instance, in arts. The labour for material substitutions and effective material use reduce the total impacts of income, whereas the labour addition increases income given impacts. The impact per material unit, i/m, shows effects of the physical material properties on environmental qualities. Even a single material use can cause several effects. This variable is usually covered by several environmental indicators, such as the pollution, water use, waste, biodiversity loss and so on.

In the empirical studies, it is usual to combine population and income per person into the national income, gross domestic product (GDP) or consumption. These indicate that material use and environmental impacts affect, per unit income, respectively output or current expenditures. The high material use per income or expenditures is called material-intensive. It is versus the material-extensive when this use is low because much labour is used. When the growth of income or expenditures is higher than the environmental impacts, this is usually labelled as decoupling. A relative decoupling is observed when the impacts per income or expenditure unit decreased though the total impact increased. An absolute decoupling is when the total impacts decrease along with more income and expenditures. The absolute decoupling reduces pressures on environmental qualities. These three types of innovations in relation to the income growth and environmental impacts are discussed after empirical testing of the IPAT model with historic data.

2.2 Decoupling Impacts and Income

Many studies have tested the IPAT model. Empirical studies indicate that the national income and environmental impacts are not as closely related as the IPAT model suggests. Doubts have emerged when it is observed that environmental impacts have increased during the industrial expansion in the 1950s and 1960s but decreased relative to income above certain total income. The turning point in the United States, it is hypothesized, would be around the statistical average USD 5000 income per person. This shift in the relation between income and environmental impacts would be due to the larger scale of production, better technology and environmental policies and would hold for many countries (Grossman and Krueger

1991). The impacts function of income growth is then presented as an inverse U curve called the 'Environmental Kuznets Curve', after Kuznets assumption about the inverse U income distribution function of income growth (inter alia, it is proved to be misconception). Empirical tests with various materials and pollution data have not shown a consistent support of this decoupling hypothesis across various environmental qualities and countries (reviewed in de Bruijn and Heintz 1999; Stern 2004). The results vary with respect to the countries' environmental policies (Dasgupta et al. 2006) and pace of substituting obsolete technologies for new ones in time and between countries (Bertinelli et al. 2012). The decoupling is observed, but the underlying factors vary too much for a consistent definition of its causes.

The global material analysis during the period 1900–2005 indicates that the decoupling is a process ongoing throughout the last hundred years and possibly even longer though the pace of decoupling varies over time, across countries and businesses. The material use is assessed. It is an approximation of environmental impacts on assumption of constant impacts per material. Trends in the global GDP measured by real monetary term and in the material use measured by mass show that the annual global GDP growth per person has been below 1.5 % average during the last century and that the average annual material use per person has increased 2–3 times slower than the income growth. When the global GDP and material use are corrected for the population growth, the income has grown 1.5–2 times faster than the material use throughout the last century. Only during a few decades after the World War II, the growth of material use was above 1 %, which is exceptionally high, but the income growth was twice higher. The global economy has continuously dematerialized. The material composition has also changed. The industry's share in the global material use has grown from 3 % in 1900 to 9 % in 2005; the construction share, construction is the most material-intensive activity, has increased from 9 % to about 40 %, and the biomass and energy shares have decreased (Krausmann et al. 2009). The relative decoupling is a trend (and the residents of fancy houses and offices can apparently be blamed for the growing environmental impacts, not the big shoppers).

Furthermore, several studies indicate an absolute decoupling of the income growth from material use and emissions in several high-income countries. The carbon dioxide emissions, which cause climate change, have decreased in the European Union, Japan and the United States this century, and if the Chinese and international transport emissions would grow less fast, the global carbon dioxide would stabilize (PBL 2013). The GDP of the European Union as a whole has grown this century along with lower total greenhouse gasses emission, lower acidification emissions and less nitrification of water, but other impacts are high or have increased, in particular the impacts on biodiversity (EEA 2013). Progress is also observed in the low-income countries whose income grows along with the decoupling, and this process evolves much faster than in the high-income countries, although many low-income countries are more material intensive than the latter (Luken and Castellos-Silveria 2011). An environmental sustainability assessment of 14 high-income countries and 11 low-income countries with reference to the Millennium Development Goals has shown no worsening in the last 20 years and much

2.2 Decoupling Impacts and Income

improvement in Brazil, Singapore and Germany. The fast progress is partly because the indices for progress in renewable energy compensate slower progress in other areas, but much progress in the high-income and low-income countries across the goals holds true after sensitivity analyses for the countries' renewable energy (Sengupta et al. 2014). These findings show that the material use and emissions decrease relative to income and in total in several countries during long periods.

The relative and absolute decoupling also holds for the private consumption value, which indicates current expenditures. This excludes the governmental expenditures, which cover many material-extensive services and several non-material financial components in the GDP accounts, such as rents and liabilities. The private consumption value, therefore, is more directly related to the material use, and environmental impacts are more sensitive to the changes in private consumption value than to the changes of GDP. It is also closer to the notion of affluence in the IPAT model than the GDP. Based on the IPAT model, it should be expected that environmental impacts of affluent societies increase along with their consumption growth. Hence, the income growth, private consumption value and material use in 34 high-income countries associated in the OECD is assessed with the statistical OECD data (www.stats.oecd.org/). Data on GDP and private consumption value are used; the GDP is used for control of results. The OECD countries are high income. The GDP of the 34 OECD members covers about 59 % of roughly USD 65 (€ 50) trillion global GDP in current value, which embraces the production value of 192 United Nations member countries. The OECD consumption is affluent from the usual environmentalist viewpoint because all these countries consume per person well above the carrying capacity of environment estimated as natural sinks to absorb environmental impacts, labelled as the ecological footprint (Wackernagel et al. 1999; WWF 2005; Global Footprint Network 2014). The OECD data covers only the domestic material extraction without data on imports and exports, which is discussed in the next paragraph. The data cover the period 1980–2010. This period embraces a large part of the ICT-based long wave after a slow down during the late 1970s, which means the recession during 1980s, revival of 1990s and decay after 2002 with the financial crisis in 2008. Coverage of the expansion and slow-down periods avoids the bias of solely low growth or high growth economy. The monetary values are in constant USD_{2005} prices. The material use is in tons. The calculations of the average growth of GDP, private consumption value and material use in that period are followed by the cross-country correlations of average income, respectively, consumption and material growth. The Pearson correlation is used because it is easy to process:

$$R^2(x,y) = \frac{\sum(x-\bar{x})\cdot(y-\bar{y})}{\sqrt{\sum(x-\bar{x})^2 \cdot \sum(y-\bar{y})^2}} \qquad (2.4)$$

where R^2 is the standard correlation, x the annual average private consumption value change in the period and y the annual average change material use in the period.

Also standard deviations of the cross-country correlations are calculated to indicate fluctuations per year. A high standard deviation means much fluctuation, which implies a weak trend. A combination of high correlation and low standard deviation means strong correlation between data sets.

$$s = \sqrt{\left[\frac{1}{(N-1)}\right] \cdot \sum_{i=1}^{N}(x_1 - \bar{x})^2} \qquad (2.5)$$

where s is the standard deviation, x the observed values and N the number of observations.

The calculations confirm the assumption that the private consumption value is closely linked to the material use than the GDP (correlation of annual average growth of the private consumption value with the material use $R^2 = 0.61$, $s = 1.32$; the GDP to material use is $R^2 = 0.50$, $s = 1.75$). Based on this data, two main findings can be pinpointed.

One finding is that in nearly all OECD countries, the private consumption values have increased faster than the material use. It also holds for the emerging economies of Brazil, China, India, Indonesia, Russian Federation and South Africa. This confirms the global assessments above though there are differences in the rate of decoupling between the countries. Out of the total 34 OECD countries, 20 countries increased the private consumption value and material use. Out of these 20 countries, only 2 countries, which are Portugal and Slovenia, increased their value slower than their material use. Hence, 18 countries reached the relative decoupling, 2 did not. Fourteen countries increased the private consumption value along with the lower total material use. These countries reached an absolute decoupling. Out of these, Poland, Slovak Republic and the United Kingdom increased their private consumption value above the OECD average. More domestic material use is not necessary for the private consumption value growth. Whether inclusion of the material imports from the low-income countries would change these results is discussed beneath.

Second finding refers to indicators for environmental impacts. The growth of private consumption value in the OECD countries is compared to the changes of six indicators of environmental impacts. These indicators are selected mainly due to data availability. These indicators address various environmental impacts: material use is a general impact indicator, greenhouse gasses indicate climate change, nitrous oxides (NOx) emissions indicate acidification and toxicity, forestry area indicates biodiversity and generated waste indicate squandering and water use for the water abstraction and pollution. The estimates are made in total and per person. The period is 1991–2010, which covers the period of economic prosperity. It is 11 years shorter than the period of the material data because more data is not available. Table 2.1 shows the OECD countries divided into increasing and decreasing impacts and into faster or slower growers than the OECD average measured by the private consumption value. All this data is per person. More than half of all OECD countries decreased their total material use, greenhouse gasses, NOx emissions and water abstraction. These countries attained an absolute decoupling. The private consumption value in all these lower-impact countries, except in one for the material use, has grown faster than the OECD average. Biodiversity and waste are difficult to reduce in most OECD

Table 2.1 OECD countries performance 1991–2010 using annual average growth of the private consumption value and environmental indicators (www.stats.oecd.org/)

Number of countries: impacts related to 2.4 % OECD growth	n	Impact on	More impacts >2.4	≤2.4	Less impacts >2.4	≤2.4	% high per-former
Material use	34	Minerals	11	2	10	11	94 %
Greenhouse gasses	34	Climate	8	4	13	9	88 %
NOx emissions	33	Health	5	0	15	13	100 %
Forestry area, non-lodging	34	Biodiversity	14	8	7	5	82 %
Waste generation	32	Recycling	13	9	6	4	68 %
Freshwater abstraction	23	Water use	4	2	10	7	96 %

countries. The countries in which the environmental impacts have decreased are generally the fast growers measured by private consumption value. The last column in the Table shows the number of countries that have reduced their impacts in total or relative and increased the private consumption value above the OECD average. These 'high performers' do well across most of the environmental indicators, as well as by income and private consumption value.

This assessment adds to the previous studies that most of the fast-growing, high-income countries measured by the private consumption value have attained the absolute decoupling with environmental impacts except for waste and biodiversity and that nearly all high-income countries have reached the relative decoupling.

2.3 Conventional Explanations

How to explain these findings? Three arguments regularly appear in the debates about the decoupling: high and increasing natural resource prices would invoke material reduction, more materials would be imported from the low-income countries and policies would enforce better environmental performance.

2.3.1 Natural Resource Prices

The increasing prices of natural resources would invoke material reduction, but observation is that the decreasing trend of real natural resource prices throughout the last 200 years was observed. John Stuart Mill already observed this in 1848: 'But the crude material generally forms so small a portion of the total cost, that any tendency which may exist to a progressive increase in that single item, is much overbalanced by the diminution continually taking place in all the other elements; to which diminution it is impossible at present to assign any limit' (Mill (1884) 1985:64). The decrease of natural resource prices continued in the last century. This is explained by the backstop technologies; it means technological alternatives whose marginal costs are lower than the resource prices and therefore being introduced. It is considered an autonomous technological change (Dasgupta and Heal 1979). It is also hypothesized that the decreasing trend of the natural resource prices would turn in the 1960s into the increasing trend due to high-income growth. The long-run price trends would be represented by parabolic lines (Slade 1982). This hypothesis is assessed incompatible with the empirical prices and is dismissed because the projected prices after 1960s based on the parabolic lines have not matched the observed prices in the 1980s. The observed prices were decreasing as a trend albeit with fluctuation (Ahrens and Sharma 1997). The high global income growth could have invoked the price increases, but this was a short period event followed by the slumping prices. When the price fluctuations are calculated as trends, the real prices of natural resources were decreasing with a few short periods of price peaks due to high oil and gas prices in the 1980s. This decreasing price trend is explained by the decreasing extraction costs due to innovations (Krautkraemer 2005). The material intensity of economies decreases despite the decreasing real prices of natural resources because the costs of technologies for extracting and using natural resources decreased faster than the material use. The prices do not explain the decoupling in the OECD countries.

2.3.2 'Outsourcing' Environmental Impacts

Another conventional explanation of the observed decoupling is that material-intensive businesses, which are usually the largest polluters, emigrate from the high-income countries to start similar activities in the low-income countries with lax environmental regulations. This way pollution control costs would be avoided. The material-intensive and polluting industries would 'outsource' environmental impacts to the low-income countries from where the manufactured products would be exported to the high-income countries. China is often mentioned as an exemplary pollution haven for the manufacturing and export of the manufactured products to many countries. No doubt that material use and pollution are huge in many low-income countries and that the manufacturing is mobile, but the argumentation does not hold in general.

One argument against the outsourcing thesis is that the effect of strict environmental regulations on emigration is disputed because these regulations rarely matter for the firms costs compared to other factors, such as cheap labour and large markets (e.g. Copeland and Scott Taylor 2003). Moreover, the OECD countries largely trade between themselves measured by weight. For instance, the European Union import from China, which is the largest exporter to Europe, was nearly 17 % of all import and from the United States nearly 12 %, but the bulky imports of primary materials were only 3 % and 19 %, respectively, all measured by value; it is based on data of the European Trade in Goods with China and the United States, respectively. The OECD and more specifically Europe have a large material-intensive industry.

Second, the international commodity trade does not confirm the outsourcing thesis. If the OECD countries would systematically import more material-intensive goods than they export, then changes of the net imports, which is imports minus export, would be reflected in changes of the domestic material use (all measured by value). However, during the period 1986–2010, there is a random fluctuation (the cross-country correlations of the average annual trade balance with the average annual material use is: $R^2 = -0.06$, $s = 162$; it is similar for the average regression with the annual cross-country correlations). A number of the OECD countries whose material use has grown have even enlarged their exports to become net material exporters. The global cross-country studies on the decoupling rates also show that the rates vary across countries and period. It means that several high and low-income countries increase income and decrease material use (Dittrich et al. 2012). Moreover, the OECD domestic construction material use has grown faster than other material uses, and the imports of these materials from the low-income countries are low. The low-income countries' exports grow, but these are not necessarily more material-intensive than the trade within the OECD.

Third, the European Union statistical data does provide material trade data during 2000–2012 (www.eurostat.eu). This was the period of economic expansion until 2008 but slowed down thereafter. Several but not all European Union countries are OECD members. This data shows that the material use in the European Union has decreased, measured by the domestic use including its trade balance by −0.7 % annual average with annual fluctuations. The GDP in this period has increased. It is an absolute decoupling albeit unimpressive.

2.3.3 Environmental Policies

It could be that the environmental policies act resolutely to reduce the impacts. The OECD data, however, put doubts about this explanation. The resolute environmental policies could be indicated by percentage of the environmental taxes in total taxes in the OECD countries because the high percentages indicate much influence of environmental issues in the countries policymaking. The correlations between the countries' share of environmental taxes in all taxes with the environmental impacts mentioned above, however, are generally low. Two kinds of correlations are made: the average volume during the period 1994–2010 across countries and annual average change of volume during 1995–2010 for the OECD total. High negative correlation would indicate influence of environmental policies on the impacts. Herewith, it is assumed that $R^2 > -0.5$ indicates large influence. In all cases of environmental impacts across the OECD countries, the higher share of environmental taxes in all taxes does go along with lower environmental impacts, but there are large differences across between the countries and huge fluctuations per year. Most correlations of the volume and changes are low ($R^2 < -0.3$). A few moderate correlations are for the greenhouse effect ($R^2 = -0.31$ for volume and $R^2 = 0.41$ for changes), which can be due to the gas and renewable energy promoting policies in several

OECD countries. The only high correlation is between the changes in material use and environmental taxes. It is presumably an indirect link related to the growing share of services, which is discussed in the next paragraph. Selection of the best performing countries by the environmental indicators does not improve the results. The environmental policies influence some emissions, in some countries, but they have low impacts on the decoupling.

2.4 Innovation and Decoupling

The relative decoupling of material use and environmental impacts from the income is observed in nearly all high-income countries and the absolute decoupling in nearly half of these countries though the decoupling rates vary per country and per impact. The decoupling processes cannot be explained by the increasing material resource prices, outsourcing of material-intensive manufacturing and more environmental policies though all these factors put together can be relevant in some cases. Another explanation is needed. The proposed explanation refers to three types of innovations that influence the material use of production and consumption. These innovations are labour for material substitution called services, more effective material uses in production called process innovations and value-adding use of labour in production and consumption called product innovations.

The substitution of labour for materials is observed as the service growth. In production of the high-income countries during 1950s and 1960s, the service value measured as turnover has grown per year 1.5–2.0 % faster than the GDP growth and only slightly above the economic growth thereafter (U.S. Department of Commerce 1996). The share of service value in the GDP of the OECD stabilized in the last few decades at about 75 % until nowadays. The composition of labour also changed. Presently, about 64 % of employed males and 82 % of employed females in the high-income countries work in services, even more in the European Union. The shares of services measured by value and labour are lower in the low-income countries but they grow faster (World Bank). In consumption, the substitution of labour for materials is indicated by the household expenditures measured in constant $€_{2005}$ in the European Union during the period 1995–2012, which covers expansion and slump. More data is not available. The total household expenditures are divided into the material-intensive products and labour-intensive services though all involve some labour and materials. The following expenditure categories are considered consumption of products: (1) food and non-alcoholic beverages; (2) alcoholic beverages, tobacco and drugs; (3) clothing and footwear; (4) housing, water, electricity, gas and other fuels; (5) furnishing, household equipment and routine household maintenance; (6) health; and (7) transport. The following are considered household expenditures on services: (8) communication, (9) recreation and culture, (10) education, (11) restaurants and hotels and (12) miscellaneous goods and services. The service in consumption in the European Union covered 33 % of the annual average household expenditures. It is lower compared to 35 % in the United States

2.4 Innovation and Decoupling

and higher than 27 % in Japan using the European statistics (with only a few data). It is much higher compared to about 22 % global household expenditures on services using the World Bank data with somewhat different consumption categories than in the European statistics and deficient countries data. In the European Union, the annual consumption of services has increased because all household expenditures have increased by 1.1 % annual average, but the share of services in consumption has decreased by −0.04 % annual average. The consumption of services could have increased during the last 50 years but stabilized during the last two decades, similarly to the share of services in production. In the European Union, therefore, the substitution of labour for materials in consumption is negligible. The labour for material substitution in the high-income countries could have been important for the decoupling throughout the last half century, but it is presently a minor factor in the decoupling in production and negligible in consumption. This substitution is presumably an important factor for the observed decoupling in the low-income countries and the emerging economies.

The decoupling in the last decades in the high-income countries, it is argued, is mainly due to the cost-saving reduction of material use entailing addition of labour that adds value to products. The hypothesis is that the cost-saving material reduction in production decreases environmental impacts and generates innovation rents. The rents are allocated for the value-adding labour, which generates income. This value allocation from materials to final products reduces environmental impacts along with income growth evolves autonomously due to innovations. The cost-saving material reduction evolves continuously due to the process innovations entailing in numerous adaptations. The adaptations aiming at more effective material use go on for decades, in many cases even centuries as observed in cases of capital goods (Rosenberg 1982) and utensils (Petrovsky 1994). In result, the material intensity of product life cycles decreased throughout the last half century measured in material use per unit cost as being illustrated on the national levels, in businesses and by product cases (Larson et al. 1986; Herman et al. 1989; Tilton 1991; Wright 1997). When the cost-savings due to material reduction exceed the costs involved in these innovations and adaptations, the innovation rents are gained, and when the innovation rents are allocated in labour that adds value in products, income growth is generated. The value-adding labour embraces engineering of capital goods, design of consumer products, marketing and so on. This labour addition evolved throughout many decades as observed by the globally growing number of knowledge workers, among these engineers and designers as measured in the science and engineering indicators. This allocation process can be considered an autonomous technological change. The material addition to products rarely adds value except for precious minerals in some specialties, such as jewellery. Moreover, high dependency on natural resources impedes income growth because the resources extracting firms pay well and attract capabilities leaving less to others (Sachs and Warner 2001), whereas the know-how involved in the extraction technologies adds value (Stijns 2005).

The global scale of this allocation is assessed. The allocation of innovation rents cannot be assessed directly based on the costs and sales of technologies because these data are unavailable. An indirect estimate is made. It is based on comparison of the GDP trend in the real USD_{2005} and the material use trend in tons in the OECD during 2000–2010 (earlier data is incomplete). The GDP trend, corrected for the labour costs to exclude the effect of changing labour costs, is compared to the material use trend. The increase of GDP compared to the increase of materials indicates the volume of allocated innovation rents (GDP divided by the labour cost index of 2000 and divided by the material use index of 2000 and multiplied by GDP). The allocation of innovation rents in the OECD has increased from USD 337 (€ 259) billion in 2001 to USD 11,146 (€ 8574) billion in 2010, which is USD 4789 (€ 3684) billion average increase per year. Since the OECD GDP is about 59 % of the USD 64,548 (€ 49,652) billion global GDP, the global income increase due to the allocation of innovation rents is average USD 8160 (€ 6698) billion. It is about 12.6 % of the global GDP, which is a sufficiently large income allocation to explain the decoupling of the global income and environmental impacts.

The decoupling during 1950s to 1970s in the high-income countries can be explained by the substitution of material-intensive manufacturing through labour-intensive services in production and consumption, which is presently relevant in the low-income countries. In the last few decades, the economic circulation in which innovation rents due to the cost-saving material reduction are allocated into the value-adding labour provides plausible explanation for the relative and absolute decoupling in the high-income countries. The cost savings are due to the faster rate of technological change than the rate of resource prices and the income growth is due to the faster rate of value-adding labour than the labour costs. The decoupling is inherent to the innovation-based income growth. Moreover, slow or moderate decoupling indicates insufficient innovation-based income growth. The innovation-based income growth is not obvious because income is often generated through exploitation of labour and environmental qualities instead of fostering know-how. Sustainable development, therefore, depends on the progress in know-how (as underpinned in the Appendix).

2.5 Conclusions

The discussion about whether income and environmental qualities are compatible started with the environmentalist argumentation that the environmental impacts increase because of population growth, affluence per person and harmful technologies. The global income growth would increase environmental impacts, but observations are that the material use per income has decreased throughout the last 100 years. Data on 34 high-income countries during the last decades also show that the countries' income and private consumption value have grown with lower material use and lower environmental impacts in total and per unit of income. Nearly half of these countries reduced the total material use and several environmental impacts, among these greenhouse gasses, NOx emissions and water abstraction. The waste and biodiversity performances are poor. The high-income growth rates often match fast decrease of the environmental impacts, except for waste and biodiversity. The decoupling of income and environmental impacts cannot be explained by higher real natural resource prices because the price trend is decreasing though with fluctuation. Imports of material-intensive production from the low-income countries compared to exports of the domestic material-intensive production are relevant in some cases but not in general. The strength of environmental policies has little influence on the decoupling.

Environmental impacts are reduced along with the income growth when labour substitutes materials, which means more services are introduced, as well as due to the process innovations that enable more cost-effective use of materials and the product innovations that add value due to additional know-how labour. The services have been a major factor in the decoupling in the high-income countries during the last half century, but it is a minor factor during the last few decades. It is still important in the low-income countries. In the high-income countries, the economic circulation in which the innovation rents of cost-saving material reduction due to process innovations are allocated into the value-adding labour due to product innovations does explain the observed decoupling. The scale of the allocation of these innovation rents is nearly 13 % of the global GDP. This volume is sufficiently large to underpin the decoupling of income and impacts. Given a demand for environmental qualities, the decoupling of environmental impacts from the income growth evolves largely autonomously due to the innovations but at a slow pace. Know-how development that enables innovations mentioned above enhances the decoupling. A major condition for such innovation-based income growth is expansion of know-how in a broad sense of the education, knowledge, skills, creativity and knowledge interaction. Vice versa, a sluggish decoupling indicates impediments for innovations when activities harmful for labour and environment are supported and when know-how is insufficient or stagnates. Policies that foster the know-how and abolish these impediments contribute to the income growth with lower environmental impacts.

Appendix

The autonomous technological change can be explained based on a loss prevention model. In the model, material inputs for a qualified output labelled as product are assumed to be used more effectively (process innovations). A novel product that is supposed to add value (product innovation) is considered a stochastic event. All non-qualified outputs are material losses or emissions that need rework with additional inputs or these are discharges that cause environmental impacts. The stochastic events imply that the number of outputs is factorial of the input number:

$$N_i = N_o! + 1 \tag{A2.1}$$

For instance, if an entrepreneur aims to attain a qualified product with three inputs (e.g. a unit of capital, labour, environment), six losses and one product are possible if six losses are not prevented. With four inputs, 24 losses and one product are possible if 24 losses are not prevented, and so on (van Leeuwen 1989). The product chance is reciprocal to the inputs number, e.g. with three inputs, the chance of a product is 1:7 and loss is 6:7.

Know-how enables to prevent losses. If the loss prevention increases linearly to the know-how costs, e.g. one know-how cost unit saves one material unit due to loss prevention, seven know-how cost units maximize the materials saving of three inputs. Given the input costs, the product, p_o, is the function of the know-how unit cost c_k, per material saving (or emission reduction), m_s, with α that relates know-how to materials. It is formally

$$p_o = \alpha \cdot c_k / m_s \qquad (A2.2)$$

An observation is that the product is a logarithmic function of know-how unit cost given material saving (or emission reduction), which can be approximated by the Poisson distribution of unit costs (it describes probability of output given limited number of stochastic events). The know-how costs increase exponentially with the material saving (Krozer 2008). It is formally

$$p_o = \alpha \cdot \ln\left(c_k / m_s\right) \qquad (A2.3)$$

in case of 3 inputs: $c_k = m_s \cdot 2.78^{7\alpha}$

The know-how on loss prevention (process innovation) can increase as long as its cost increase is lower than the material cost savings. The lower cost generates innovation rent, c_s. A novel product (product innovation) is introduced if labour costs, c_v, are below the rent and add value, a_v.

Given is

$$I_t = P \cdot a/p \cdot m/a \cdot i/m \qquad (A2.4)$$

Herewith, the consumption (affluence), a, increases along with lower environmental impacts, i, if innovation rents of the process innovations are allocated for the value-adding product innovations, v_a, and the material saving due to process innovations, m_s, is larger than the material addition due to products innovation m_v, formally at $t+1$:

$$I_{t+1} = P \cdot (a+v_a)/p \cdot (m-m_s+m_v)/a \cdot i/m \qquad (A2.5)$$

$$I_{t+1}/I_t = \frac{P \cdot (a+v_a)/p \cdot (m-m_s+m_v)/a \cdot i/m}{P \cdot a/p \cdot m/a \cdot i/m} \qquad (A2.6)$$

$$i = \frac{v_a}{a} \cdot \frac{(-m_s + m_v)}{p} \qquad (A2.7)$$

$$V_t = P \cdot a \qquad (A2.8)$$

$$V_{t+1} = P \cdot \left[v_a - (c_s + c_v) \right] \qquad (A2.9)$$

$$\frac{V_{t+1}}{V_t} = v = \frac{\left[v_a - (c_s + c_v) \right]}{a} \qquad (A2.10)$$

If

$$i < 0 \text{ if } m_s > m_v$$
$$v = v_a \geq c_c + c_v$$

Incomes grow along with lower environmental impacts if (a) material saving due to process innovations is larger that additional material use for value addition, which is observed, and (b) know-how costs needed for the material saving and value addition are below the value addition, which is plausible in competition. Sustainable development depends on the costs and effects of know-how involved in material saving and value addition.

References

Ahrens, W. A., & Sharma, V. R. (1997). Trends in natural resource commodity prices: Deterministic or stochastic? *Journal of Environmental Economics and Management, 33*, 59–74.
Alcott, B. (2010). Impact caps: Why population, affluence and technology strategies should be abandoned. *Journal of Cleaner Production, 18*, 552–560.
Bertinelli, L., Strobl, E., & Zou, B. (2012). Sustainable economic development: Theory and evidence. *Energy Economics, 34*, 1105–1114.
Chertow, M. R. (2001). The IPAT equation and its variants. *Journal of Industrial Ecology, 4*(4), 13–29.
Copeland, B. R., & Scott Taylor, M. (2003). *Trade and the environment* (1st ed.). Princeton: Princeton University Press.
Dasgupta, P. S., & Heal, G. M. (1979). *Economic theory and exhaustible resources* (1st ed., pp. 173–192). Cambridge: Cambridge University Press.
Dasgupta, S., Hamilton, K., Pandey, K. D., & Wheeler, D. (2006). Environment during growth: Accounting for governance and vulnerability. *World Development, 34*(9), 1597–1611.
de Bruijn, S. M., & Heintz, R. J. (1999). The environmental Kuznets curve hypothesis. In J. C. J. M. van den Bergh (Ed.), *Handbook of environmental and resource economics* (1st ed., pp. 656–677). Northamptonm: Edward Elgar.
Dittrich, M., Giljum, S., Lutter, S., & Polzin, Ch. (2012). *Green economies around the world? Implications of resource use for development and the environment*. Vienna: SERI, Mimeo.
EEA. (2013). *Environmental pressures from European consumption and production* (Technical Report 2-2013). Copenhagen: European Environmental Agency.
Ehrlich, P. R., & Holdren, J. P. (1971). Impact of population growth. *Science, 171*(3977), 1212–1217.

Global Footprint Network. (2014). http://www.footprintnetwork.org/en/index.php/GFN/page/at_a_glance/. Visit 20 July 2014.
Grossman, G. M., & Krueger, A. B. (1991). *Environmental impacts of a North American free trade agreement* (Working Paper Series, 3914). NBER, Mimeo.
Herman, R., Ardekanin, S. A., & Ausubel, J. H. (1989). Dematerialization. In J. H. Ausubel & H. E. Sladovich (Eds.), *Technology and environment* (1st ed., pp. 50–69). Washington, DC: National Academy Press.
Krausmann, F., Gingrich, S., Eisenmenger, N., Erb, K.-H., Haberl, H., & Fischer-Kowalski, M. (2009). Growth in global materials use, GDP and population during the 20th century. *Ecological Economics, 68*, 2696–2705.
Krautkraemer, J. A. (2005). *Economics of natural scarcity: The state of the debate*. Washington, DC: Resources for the Future, Mimeo.
Krozer, Y. (2008). *Innovations and the environment*. London: Springer.
Larson, E. D., Ross, M. H., & Williams, R. H. (1986). Beyond the era of materials. *Scientific American, 34*(6), 24–29.
Luken, R., & Castellianos-Silveria, F. (2011). Industrial transformation and sustainable development in developing countries. *Sustainable Development, 19*(3), 167–175.
Mill, J. S. ((1884) 1985). *The principles of political economy* (Book IV, Chap. VI). New York: Penguin Classics.
PBL [Plan Bureau voor Lefomgeving]. (2013). *Global CO2 emissions from fossil fuel use and cement production per region, 1990–2012*. Bilthoven, Mimeo.
Petrovsky, H. (1994). *The evolution of useful things* (1st ed.). New York: Vintage Books/Random House.
Rosenberg, N. (1982). Learning by using. In N. Rosenberg (Ed.), *Inside the black box* (pp. 120–140). Cambridge: Cambridge University Press.
Sachs, J. D., & Warner, A. M. (2001). Natural resources and economic development: The curse of natural resources. *European Economic Review, 45*, 827–838.
Sengupta, D., Mukherjee, R., & Sikdar, S. K. (2014). Environmental sustainability of countries using UN MDG indicators by multivariate statistical methods. *Environmental Progress & Sustainable Energy, 34*(1), 198–206. doi:10.1002/ep.11963.
Slade, M. (1982). Trends in natural-resource commodity prices: An analysis of the time domain. *Journal of Environmental Economics and Management, 9*, 122–137.
Stern, D. I. (2004). Rise and fall of the environmental Kuznets curve. *World Development, 32*(8), 1419–1439.
Stijns, J.-P. (2005). Natural resource abundance and economic growth revisited. *Resources Policy, 30*, 107–130.
Tilton, J. E. (1991). Material substitution: The role of new technology. In N. Nakicenovic & A. Grubler (Eds.), *Diffusion of technologies and social behaviour* (pp. 383–406). Berlin: Springer.
U.S. Department of Commerce. (1996). *Economics and statistics administration*. Washington, DC: Office of Policy Development, Service Industries and Economic Performance.
van Leeuwen, C. (1989). De organisatie van milieu en veiligheid in een grote onderneming. In H. Vollebergh (Ed.), *Milieu en Innovatie* (1st ed., pp. 155–176). Groningen: Wolters Noordhoff.
Wackernagel, M., Onisto, L., Bello, P., Linares, A. C., Lopez Falfan, I. S., Mendez Garcia, J., Suarez Guerrero, A. I., Ma, G., & Guererro, S. (1999). National natural capital accounting with the ecological footprint concept. *Ecological Economics, 29*, 375–390.
Wright, G. (1997). Towards a more historical approach to technological change'. *The Economic Journal, 107*, 1560–1566.
World Wildlife Fund. (2005). *Ecological footprint Europe 2005*. Brussels: WWF European Policy Office.

Chapter 3
Markets of Sustainable Innovations

How are sustainable innovations induced? Possibilities of induced sustainable innovations are discussed. Opportunities are due to private and social demands for environmental qualities. The demands increase because the growing knowledge work and leisure time need high environmental and cultural qualities to perform. The traditional demands refer to material resources use and pollution controls. The emerging demands address attributes of environmental qualities in man-made products. These natural blends cover ethical purchases, ecosystem services and cultural expressions. All demands together generate the global market of sustainable innovations that exceeds USD 2980 billion. It is more than 4.6 % of the global GDP. Impediments refer solely to the policy financial support of the vested interests that are rivals to sustainable innovations: tax exemptions and subsidies for fossil fuels, support of environmentally harmful agricultural practices, unnecessary infrastructure and concessions for resource extraction. These exceed USD 3053 billion. The social demands are sufficient for sustainable development, but policies impede progress.

3.1 Inducing Innovations

When innovation-rents due to the cost-saving material reduction are used for the value-adding labour, incomes grow and environmental impacts decrease. Given innovations, this process evolves largely spontaneously, but the pace is slow compared to aspirations for sustainable development. Faster progress is needed. Inducing sustainable innovations is discussed with regard to opportunities due to the private and social demands for environmental qualities and impediments caused by the policy support of environmentally harmful activities. The demands are expressed as possibilities of sales of sustainable innovations, the impediments as the financial support of the vested interests that are rivals to the sustainable innovators. The assumption is that sustainable innovations are pursued if the innovators expect many possibilities and few barriers. Under this condition, environmental impacts are reduced along with income growth. The pace of impact reduction, Δi, is a function of the allocated innovation-rents, Δv_a, subject to demands for environmental qualities, d, versus barriers for innovations, b, t time; formally:

$$\Delta i = \beta \Delta v_a^{(d/b)t} \qquad (3.1)$$

Given is β, a coefficient that relates autonomous decoupling of income growth from environmental impacts: $\beta = f(m/a)$ as discussed in Sect. 2.2. It is shown that this coefficient is usually below one across countries, periods and indicators.

3.2 Demands for Sustainable Innovations

The high and growing demands for environmental qualities are observed in nearly all high-income countries. These demands are expressed as being willingness to pay for good environment, preferences for ethical products, active citizenship in favour of environment, membership of environmental organisations and suchlike personal preferences (Stern 2000; Viñuales 2013). The reason for these demands, which is argued, is the growing knowledge work. Artists, scientists, engineers, educators, managers, policymakers and other experts and craftsmen enlarged throughout the last century from a group of specialists to a quarter of all waged and non-waged labour in the present high-income countries. The low-income countries are busy to catch up fast. These professionals are usually the best paid ones, which attracts high human capabilities. The capabilities generate influence on political decisions, which reinforces improvements of the knowledge workers' income, working and living (Drucker 1993). The growing knowledge workers also matches the lower environmental impacts (e.g. in the European Union, the cross-countries correlation of materials used for scientific personnel during 1981–2012 is R^2 −0.28, $s=0.16$).

The interest of the knowledge workers in good environment is not solely due to the personal preferences mentioned above because these preferences can be found across all social groups. The knowledge workers have also a professional interest in environmental qualities. The interest is that the knowledge workers perform poorly in bulky situations, such as factories, but flourish when social interactions generate the knowledge spillovers. These interactions need the diversity of people, culture and high environmental qualities, and this is found in public spaces. The densely populated public spaces of cities attract knowledge workers because these enable interactions that generate knowledge spillovers (Florida 2002), but spaciousness, tranquillity, beauty and other environmental and cultural amenities are necessary for the knowledge work. These amenities are found outside the cities, and they are recreated in the urban context as parks, lawns, campuses and suchlike. These are cities' valuable assets. The demands for environmental qualities also grow due to the growing leisure time because of the leisure benefits from good environment. Income growth invoked shorter work time along with longer lifetime. Between 1990 and

2010, the annual average working hours have shortened by 0.4 % in the high-income countries; it is 1.8 % below the average real income growth in that period, and the average lifetime has elongated by 0.3 % a year (OECD data). Compounded for 20 years, the leisure time has increased on average per person by 20 %, whereas the real average income per person by nearly 100 %. More leisure time and even higher income growth enable to spend more on environmental qualities for leisure.

The traditional demands for environmental qualities address availability of fuels and minerals, health and safety issues and suchlike. These demands, which are often expressed by authorities as policy, generate businesses aiming to satisfy these policy demands. These are businesses in renewable energy, water and waste called 'cleantech'. The private investors' data suggest that the global investments in the cleantech in 2010 were USD 499 (€372) billion, out of it 45 % in renewable energy, 14 % in energy efficiency, 34 % in water supply and wastewater treatment, 5 % waste treatment and 2 % others (Copenhagen Cleantech Cluster 2012). The markets are much larger when technologies for exploitation of natural resources and pollution control are considered, which is done beneath. In addition, people demand use of good environment. The social demands invoked businesses that use various environmental qualities for products and services. An example is the international tourism service that has grown from close to nil arrivals and sales in the early 1960s to about 800 million arrivals and nearly USD 1100 billion sales in 2010, as measured statistically by the World Bank and World Tourism Organization. Most destinations are sunny seaside because hot seaside with cool breeze from the sea is a key environmental quality for leisure, but this climate is scarcely available on the global scale; no wonder that most destinations are packed. The domestic and local leisure sites are visited more frequently, and these visits generate higher incomes than the destinations far away. Moreover, the knowledge workers are mobile. The growing physical and virtual mobility due to transportation and media services has generated a globalised so called 'experience' economy and dissemination of communication tools for such services enable to sense an environmental quality and respond to its degradation nearly instantly. The demands for environmental qualities during work and leisure and the communication possibilities have generated markets for sustainable innovations, which add to the markets of material resource and pollution management.

The consumptive uses of environmental qualities emerge. The individual consumers use blends of culture and nature, referred to as the 'natural blends'. These are man-made, attributes for private uses mixed with the collective uses of environmental qualities (such mixes are common, e.g. marriages coexist with markets of love and public armies with private security forces). The natural blends can be divided into three markets. The first market covers consumption of the environmental quality attributes in products, for instance, the 'natural materials', 'organic food' and 'green banks'. The second one embraces personal interactions with environmental qualities, e.g. visits to nature parks, bird watching and tree-hugging. The third market covers cultural expressions of environmental qualities in rituals and media, arts and crafts, science and education and so on. These natural blend markets allow for various services, and businesses emerge assumed intangible one generation back, be it cuddling orangutans, the Jungfrau panorama or space tourism. Many services are controversial and need sustainable innovations.

The traditional and emerging markets of sustainable innovations are assessed. The present technology uses are considered as being the reference market. The total value assessment is used as framework for the estimates. This framework covers the direct and indirect user values, as well as the non-user values which cover option values, bequest values and existence values (McNeely 1988; Hanley 1992). The user values are estimated using prices and volumes based on statistical data. The non-user values have no prices. These can be estimated using willingness to pay for environmental qualities compared to alternative qualities. Such contingent valuation methods, which can be found in many handbooks, show preferences that reflect the prevailing social conventions and morality rather than the people's readiness to spend on environmental qualities (Kahneman and Knetsch 1992). Therefore, opportunities for sales of sustainable innovations are approximated based on scaling of the present markets instead of estimating willingness to pay. Tourism is excluded to avoid double counting.

The market that refers to the direct user values addresses natural resources with market prices, for example, drinking water. The costs of extracting natural resources are used as sales opportunities for sustainable innovations. The market that refers to the indirect user values covers the pollution control technologies and management. Pollution is an external effect that has no market prices but quasi-prices imposed by authorities and stakeholders, e.g. effluent fees on waste water. The sales opportunities are assessed based on the public and private pollution control expenditures. These two markets are estimated with statistical data. The markets of the non-user values address the natural blends. Statistical data or quasi-prices are unavailable. The option values cover attributes of environmental qualities in products, for instance, sales of water with health claims. The sales opportunities are estimated based on the present sales of products with such attributes called ethical consumption. The market of bequest values covers benefits of environmental qualities for future generations, for example, biodiversity of a lake. The sales opportunities are based on the concept of using environmental qualities called ecosystem services (though it reflects the anthropocentric thinking rather than service). Scaling up of the nature park management is used for approximation of this market. Finally, the existing values are cultural expressions of environment, e.g. a lake as an inspiration for paintings. This market is estimated with the expenditures of environmental activists.

3.3 Markets of Sustainable Innovations

Sales opportunities of sustainable innovations are assumed when private and social demands for environmental qualities are tangible and sustainable innovations can outperform the available technologies measured by costs and quality. These are considered with respect to natural resources (material resources), external effects (pollution reduction) and natural blends (ethical consumption, ecosystem services and cultural attributes). All estimates are crude aiming to indicate the scale of the global sales opportunities. The markets are expressed in USD, Euro and as the percentage of the USD 64,548 (€ 49,652) billion global GDP in 2010 (more recent data is deficient).

3.3.1 Material Resources

The material-reducing process innovations within firms are assumed autonomous because the firms' internal operations evolve largely independent of the demands for environmental qualities except when they cause hazards for labour. Contrary to the internal business operations, land use for extraction and cultivation of the natural resources is sensitive to these demands because various interests claim land. Data on technologies and costs of the land use, however, is not available. Hence, the market of sustainable innovation is assessed indirectly. One method to assess this is the World Bank data on rents generated from extraction of natural resources; rents are sales minus the extraction costs. The data is on coal, forests, minerals, gas and oil extraction. The drawback of this data is that the additional rents do not necessarily indicate technological progress but reflect changes of the commodity prices as well as the costs of technologies but fluctuations of the commodity prices are too large for assessing these costs. Another data is used. It is the adjusted savings of natural resource depletion, i.e. comparison of the annual depletion to reserves in monetary value terms. Although this data is also deficient because this reflects expectations about the future value of natural resource, which is imperfect and can be manipulated, the larger savings indicate technological progress in development and extraction. The savings during 2004–2012 are given. The annual average increase was USD 5800 (€ 4.400) billion, but this data show strong fluctuations after financial crisis in 2008 possibly due to the changing expectations. A prudent estimate is during 2004–2008 when the annual average increase of the savings was USD 1063 (€ 818) billion. It is 1.65 % of the global GDP.

3.3.2 Pollution Control

The second market addresses technologies and management for pollution prevention and reduction, abbreviated as pollution controls. The reference market is based on the annual capital and operational costs of control of all kinds of pollution and hazards. Global statistical data on these expenditures is not available. The OECD data for the period 1997–2010 suggests that the annual average expenditures are USD 124 (€ 95) billion, but this does not cover several large countries, such as Australia, Canada, Japan and the United States, and the data on a few countries is deficient, e.g. on Mexico. The European Union statistical data is comprehensive for the period 2002–2010. This data shows the annual average costs of USD 280 (€ 216) billion. The data covers the European Union of 28 member states. These costs are in average 1.48 % of the GDP, which fluctuate between 1.4 and 1.6 % of the GDP (Georgescu and Cabeça 2010 show 1.8 % for 25 member states in 2006). Although the pollution control costs as percentage of GDP vary between 0.5 and 2.5 % across countries, fluctuate in time and depend on definitions, the European Union average is illustrative for many situations because it includes the high-income and low-income countries. The global market of sustainable innovations for pollution controls is based on scaling up of the European Union expenditures in the GDP. Using 28.5 % average European Union GDP in the global GDP based on the World Bank data, this global market is about USD 973 (€ 748) billion a year, i.e. 1.48 % of the global GDP.

3.3.3 Ethical Consumption

The option value of consumption is the ethical consumption. This is represented by purchases and uses of products and services with labels certificated by authorities. The labels indicate good working conditions, trade relation, environmental and health performance and so on. The ethical consumption can be found in businesses, finance, government and households. Herewith, only the households are covered to avoid double counting with the market of material resources and pollution control, but it underestimates the market of sustainable innovation for the ethical consumption. Statistical data on the consumer ethical expenditures is unavailable, and the available data is deficient, e.g. the European Union statistics show that 1067 products are licensed with the Ecolabel, but there is no sales data. Verified studies are also scarce. Statistical data on the consumer ethical expenditures in the United Kingdom cover purchases of foods and drinks, home-related products, travelling and transport, personal care products, community expenditures and ethical money. In the year 2011, a total of about 47 billion British pound (£) is spent (€ 59, USD 77 billion). These expenditures have grown between 2002 and 2011 on average about 4.7 % a year. It is per household average of £ 989 (€ 1251, USD 1627), which is about 3.1 % of all expenditures per person. A higher share of the food ethical consumption is observed: Austria 6 %, Denmark 5.1 %, Switzerland 4.7 %, Germany 3.4 but the Netherlands 2.3 % (Bunte et al. 2010). A prudent assumption about the expenditures on ethical consumption is 2 % of all disposable consumer expenditures in the OECD countries and 1 % in the non-OECD countries. The disposable consumers' incomes in 2010 were USD 27.1 trillion and 11.7 trillion, respectively. Hence, the market of sustainable innovations for the household ethical consumption is about USD 658 (€ 506) billion. The ethical consumption covers about 1.02 % of the global GDP.

3.3.4 Ecosystem Services

The market assessment of ecosystem services refers to the concept of nature as being provision of services for human's activities and supporting ecological activities that can be used by humans and nature being divided into marine, coastal and terrestrial areas (biomes), each area with several ecosystems (Reid 2005). The ecosystem services would be provisions (e.g. food, genes), regulations (e.g. purification, pest control) and culture (e.g. spiritual, science). Their global value, which is assumed to approximate the value of nature resource for human activities, is estimated to be USD 16,000–54,000 billion a year (Constanza et al. 1997). With regard to the global 149×10^{12} square metre land area, out of it, 38 % is cultivated (World Bank), and the assumed nature value is maximum USD 0.96 per square metre cultivated land (a remarkably low figure compared to hundreds of dollars in the urban areas). The expenditures on protection of nature are far below this figure. This land use for the benefit of future generations covers about 971×10^9 square metres mainly in the national parks, which is 1.7 % of all cultivated land. The expenditures on the nature protection areas are about USD 8.5 billion a year mainly from the public sources (Pearce 2007). The expenditure per square metre nature protection area is average USD 0.0087, i.e. 110 times lower than the minimum value of ecosystem services. Regarding the small nature protection areas and low budgets, payments for ecosystem services on the cultivated land are advocated to maintain various services. A few countries introduced such payments as fees to farmers, for instance, Costa Rica and Switzerland (Gómez-Baggethun et al. 2010), and natural parks introduce income generating activities (Lordkipanidze et al. 2012). The payment for ecosystem services on all cultivated land indicates a market of sustainable innovations. If to assume that the 0.87 dollar cent per square metre expenditure on the nature protection reflects the social valuation of ecosystem services on all cultivated land of 56.6×10^{12} square metres, the market of ecosystem services covers USD 245 (€ 188) billion, or 0.38 % of the global GDP.

3.3.5 Cultural Attributes

The value of the environment for arts, philosophy, religion, science, spirituality and other cultural attributes was the cornerstone in the environmentalist thinking throughout the last two centuries (Grober 2010). In economic theory, it is expressed as the existence value of environmental qualities. The argumentation is that pricing of environmental qualities is biased by the preferences of present generation that have deficient view on the preference of future generations and it fails to recognise that environmental qualities are valuables in themselves justifiable without recognised utilities (Krutilla 1967). Entitlements for environmental qualities are proposed, e.g. the amenity rights (Mishan 1968, 1993) and for the quality of life

(Ehrenfeld 2008). The cultural values derived from environmental qualities would foster creativity entailing better work performance (Laszlo et al. 2012) and support change management (Miller Perkins 2012). These arguments are not valuated. A minimum value of the cultural attributes based on environmental qualities is the membership of nature and environmental organisations. The minimum estimate, herewith, is 30 million members, out of them 16 millions in Europe and 14 million in the United States. The monetary value of these cultural attributes is assessed on the assumption that all these members spend on their recreation, culture, education and leisure in relation to environmental qualities. These expenditures are average 21 % of the total household expenditures in Europe and 19 % in the United States (Eurostat data). Given the average disposable income per person of USD 5657 (€ 4351) in the European Union and USD 8584 (€ 6603) in the United States, the minimum global market of the cultural attributes related to environmental qualities is USD 42.1 (€ 32.4) billion, i.e. 0.07 % of the global GDP.

3.3.6 Summary

The global market opportunities for sustainable innovations due to the private and social demands for environmental qualities are summarised in Table 3.1. The markets are added to each other because double counting is avoided and several large markets are not included, e.g. in tourism. The total market is estimated to be about USD 2981 (€ 2293) billion. It is 4.62 % of the global GDP using the 2010 year data. For a comparison, the global expenditures on health were USD 7322 billion, i.e. 10.3 % of the GDP in 2010 (World Bank), which means that the market due to demands for environmental qualities is nearly half as large as the market of health. There are opportunities for sustainable innovations.

3.4 Impediments for Sustainable Innovations

Why the large markets of sustainable innovations did not induce many more innovations is addressed. Various impediments are relevant. The impediments could be caused by the limited knowledge about environmental qualities, imperfect know-how in addressing new issues, deficient expression of the private and public demands for good environment and others. Furthermore, the impediment could be caused by policies that protect the vested interests from the creative destruction of sustainable innovations. Herewith, this scope is narrowed down to the policy financial support of the vested interests that are plausibly harmful to environmental qualities and are rivals to the sustainable innovators, and within this scope, only the well-measurable financial support is indicated. This policy impedes welfare growth and undermines fair competition because finance rivals of sustainable innovators based on taxes are imposed on all interests. Contrary to a widespread assumption

3.4 Impediments for Sustainable Innovations

Table 3.1 Global markets of sustainable innovations due to demands for environmental qualities

Markets	Description	USD billion	Share in global GDP
Material resources	Technologies for reducing material and energy use	1063	1.65
Pollution control	Cleaner technologies for pollution reduction	973	1.51
Ethical consumption	Ethical purchase and use of products and service	658	1.02
Ecosystem services	Nature management due to payment for ecosystem services	245	0.38
Cultural attributes	Cultural expression of environmental quality	42	0.07
Total		2981	4.62

that policies generally foster environmental qualities being a collective good, the present policies usually impede entry of the sustainable innovators through the financial support of their rivals.

The financial policy support of the vested interests is encountered in many businesses. It is often observed that private interests seek entitlements from authority because such rent-seeking behaviour generates monetary gains compared to competition with rivals. The social costs, however, increase when the monopoly political power is executed in a discriminatory way through entitlements for the rent-seeking interests because more costly or inferior quality solutions can be pursued compared to competition (in addition to institutional deficiencies such as corruption). The entitlements cause welfare losses (Krueger 1974). Innovators in particular are in disadvantage because being entrants on markets they have less monetary capacities and political influence than the vested interests and in addition, they need more risky investments when pursuing new solutions. The policy of awarding the rent-seeking behaviour creates barriers to entry for the innovators, which ultimately deplores innovations (Murphy et al. 1993). Such barriers of entry can persist during many decades despite social and environmental harms. For example, monopoly entitlement given to the gas producers for street lights with gas in late 1800s inhibited electric lighting during several decades despite fires caused by the gas lights (DiLorenzo 1996). Various types of entitlements create the barriers of entry. The non-monetary barriers of entry are caused by, for example, trade and work licences, property rights, concession for environmental resource, monopoly land use rights, patents and authors' rights, and the monetary ones are, for instance, subsidies, tax exemptions, credit facilities bonds and so on (Boldrin and Levine 2004). Fail-soft agreements between businesses and authorities, as well as between authorities, hinder innovators when the agreements promote undemanding solutions and obscure qualified performances. Agreements between businesses are much studied and such agreements are often persecuted. Herewith, public expenditures that reduce costs of the environmentally harmful interests rival to the sustainable innovators are addressed, and within these, a few categories are indicated. The purpose is to underpin scale of such impediments.

There are tax exemptions and subsidies for the energy production and use. This production and use are major sources of land degradation and pollution. The global policy support approximates USD 1900 (€ 1461), which is assessed by the International Monetary Fund (Clemens 2013). This support impedes innovations in energy-efficient technologies and services. Abour USD 1600 billion is support of fossil fuels including nuclear energy, which impedes greenhouse gasses mitigation, energy-efficiency and the renewable energy development. Policy interventions through taxes and subsidies on energy markets in the European Union are discussed in detail in Chap. 11.

Large subsidies are provided to the intensive agricultural businesses in the European Union, the United States and Japan. They are environmentally harmful because they focus on the enlargement of output instead of reducing social and environmental impacts of the intensive practices. The agricultural subsidies are estimated to exceed in total of USD 430 (€ 330) billion based on the European Union, the United States and Japanese data (Worldwatch 2014). Comprehensive global assessments are not found. The tax exemptions on foods, such as low VAT in the high-income countries, and subsidies for food sales in the low-income countries add to the bias in favour of output, but this global financial data is not found. The policy interventions when biased by the output orientation impede entry of the agricultural practices more in balance with fair social conditions and environmental qualities and may obstruct more healthy food consumption.

Extraction of natural resources involves much land use. When entitlements for the land use with pre-existing and customary land rights are delivered to businesses as concessions, firms are exempted from liability for sustaining environmental qualities and community life on the areas as long as they comply with the national regulations. In effect, communities are usually undermined and environmental qualities deplored. Such concessions given by authorities to firms assessed in 22 emerging economies cover nine million hectares of land at a value of USD 5190 billion (de Leon et al. 2013). Assuming on average 10 % linear depreciation of these rights, the annual value of these entitlements is about USD 519 billion a year.

Much land is used for infrastructure aiming to create roads, dams, rails and so on. The infrastructure transforms environmental qualities on-site and generates subsequent activities, which cause environmental impacts that are usually irreversible. These investments are largely based on the public funding; the global investment is estimated to be about USD 2600 billion a year, and out of it, nearly 97 % is covered by public (Dobbs et al. 2013). The public funding of infrastructure reduces risks of private investors because policies cover the costs of uneconomic works. In results, welfare losses are experienced when infrastructural works cause costs overrun. Presently, most works are not economic by the private costs and benefits, and many are not needed because they generate a negative social cost – benefit. Most of the infrastructure would not be built if based on competition because costs and risks are

Table 3.2 Impediments for sustainable innovations expressed in monetary terms

Type of impediment	USD billion
Support of fossil fuels	1600
Agricultural subsidies, excluding food	430
Tax exemptions and subsidies on foods	P.M.
Concessions for resource extraction	519
Public infrastructure	504
Value of patents	P.M.
Total	3053

too high (Flyvbjerg et al. 2003). Regarding the costs overrun, a conservative guess is that 20 % of the policy support of infrastructure is avoidable if the rent-seeking is tempered and it is socially beneficial. It is USD 504 billion a year.

The patents are entitlements to protect interests of innovators. Payments for royalties of the intellectual property rights were about USD 250 billion in 2012 (World Bank). These payments can impede the socially and environmentally beneficial innovative development, but patents are not per se supported by policies, and it is not clear which ones are harmful. These are pro memory.

The global impediments for sustainable innovations caused by the policy financial support of the vested interests are shown in the Table 3.2. The monetary value of the impediments is about USD 3053 (€ 2309) billion a year. It is 4.73 % of the global GDP. This assessment indicates that the impediments for sustainable innovations posed by policies are larger than the markets of sustainable innovations induced by private and social demands for environmental qualities and not all impediments caused by such policy interventions are included.

3.5 Conclusions

The induced technological change is addressed in addition to the autonomous technological change discussed in Chap. 2. Demands for environmental qualities create opportunities for sustainable innovators, but they must also overcome impediments. The innovators' sales possibilities can be expressed a balance of the demands and impediments. The demands grow because the growing knowledge work and leisure time need environmental qualities. They generate markets of sustainable innovations: material and energy saving, pollution control, ethical purchases, ecosystem services and cultural expressions. The total global market is estimated to be about USD 2981 (€ 2293) billion a year, which is 4.62 % of the global GDP. The impediments due to the policy financial support of the environmentally harmful vested interests impede entry of the sustainable innovators. The global policy financial support of the environmentally harmful energy use and agricultural practices, unnecessary infrastructure and environment degrading concessions is about USD 3053 (€ 2349) billion a year; and there are more policy interventions that impede sustainable innovations.

The present balance of the demand to impediment is 0.98. Hence, the present policies impede sustainable innovations, which would foster progress toward sustainable development in addition to the autonomous technological change due to the material-saving process innovations and value-adding product innovations, which is addressed in Chap. 2. For example, the abolishment of fossil fuels and agricultural support would shift the demand to impediment ratio to 2.9, which could generate 7.5 % lower environmental impacts a year along with the global 2 % income growth. This would reduce many environmental impacts to one quarter of the present ones within one generation. There is reason for optimism about income growth and good environmental quality within one generation if policies would support the sustainable innovators instead of the vested interests.

References

Boldrin, M., & Levine, D. K. (2004). Rent-seeking and innovation. *Journal of Monetary Economics, 51*(1), 127–160.
Bunte, F. H. J., van Galen, M. A., Erno Kuiper, W., & Tacken, G. (2010). Limits to growth in organic sales. *De Economist, 158*(4), 387–410.
Clemens, B. (2013). *Energy subsidies reform: Lessons and implications.* International Monetary Fund, Mimeo.
Constanza, R., d'Arge, R., de Groot, R., Farberk, S., Grasso, M., Hannon, B., Limburg, K., Naeem, S., O'Neill, R. V., Paruelo, J., Raskin, R. G., Sutton, P., & van den Belt, M. (1997). The value of the world's ecosystem services and natural capital. *Nature, 387*(15), 253–260.
Copenhagen Cleantech Cluster. (2012). *The Global Cleantech Report 2012.* Danish Industry Foundation, Mimeo.
de Leon, R., Garcia, T., Kummel, G., Munden, L., Murday, S., & Pradela, L. (2013). *Global capital, local concessions: A data-driven examination of land tenure risk and industrial concessions in emerging market economies.* The Munden Project Ltd., Mimeo.
DiLorenzo, T. J. (1996). The myth of natural monopoly. *The Review of Austrian Economics, 9*(2), 43–58.
Dobbs, R., Pohl, H., Lin, D., Mischke, J., Garemo, N., Hexter, J., Matzinger, S., Palter, R., & Nanavatty, R. (2013). *Infrastructure productivity, How to save $ 1 trillion a year.* McKinsey, Mimeo.
Drucker, P. F. (1993). *The post-capitalist society* (1st ed.). New York: HarperCollins.
Ehrenfeld, J. (2008). *Sustainability by design* (1st ed.). New Haven: Yale University Press.
Florida, R. (2002). *The rise of the creative class* (1st ed.). New York: Basic books.
Flyvbjerg, B., Bruzelius, N., & Rothegaster, W. (2003). *Megaprojects and risks* (1st ed.). Cambridge: Cambridge University Press.
Georgescu, M. A., & Cabeça, J. C. (2010). *Statistic in focus: Environment and energy.* Luxembourg: Eurostat.
Gómez-Baggethun, E., de Groot, R., Lomas, P. L., & Montes, C. (2010). The history of ecosystem services in economic theory and practices: From early notions to markets and payment schemes. *Ecological Economics, 69*, 1209–1218.
Grober, U. (2010). *Sustainability: A cultural history* (1st ed.). Padstow: T J International.
Hanley, N. (1992). Are there environmental limits to cost benefit analysis? *Environmental and Resource Economics, 2*, 33–59.
Kahneman, D., & Knetsch, J. L. (1992). Valuing goods: The purchase of moral satisfaction. *Journal of Environmental Economics and Management, 22*, 57–70.

References

Krueger, A. O. (1974). The political economy of the rent-seeking society. *The American Economic Review, 64*(3), 291–303.

Krutilla, J. V. (1967). Conservation reconsidered. *The American Economic Review, 57*(4), 777–786.

Laszlo, C., Brown, J. S., Sherman, D., Barros, I., Boland, B., Ehrenfeld, J., Gorham, M., Robson, L., Saillant, R., & Werder, P. (2012). Flourishing: A vision for business and world. *Journal of Corporate Citizenship, 46*, 31–51.

Lordkipanidze, M., Krozer, Y., Lovett, J., & Christensen-Redzepovic, E. (2012). A challenge of balancing conservation with economic opportunities in National Parks. In A. Morvillo (Ed.), *Proceedings of the 1st enlightening tourism conference 2012*, 13–14th September 2012, Naples.

McNeely, J. (1988). *Economics and biological diversity* (1st ed.). Gland: International Union for Conservation of Nature.

Miller Perkins, K. (2012). Sustainability and innovation: Creating change that engages the workforce. *Journal of Corporate Citizenship, 12*, 175–187.

Mishan, E. J. ((1968) 1993). *The cost of economic growth* (1st ed., pp. 30–46). London: Weidenfeld and Nicholson.

Murphy, K. M., Shleifer, A., & Vishny, R. W. (1993). Why is rent-seeking so costly to growth? *The American Economic Review, 83*(2), 409–414.

OECD data. http://stats.oecd.org/.

Pearce, D. (2007). Do we really care about biodiversity. *Environmental and Resource Economics, 37*, 313–333.

Reid, W. (2005). Millennium Ecosystem Assessment. 2005. *Ecosystems and human well-being: Synthesis*. Bahrain: United Nations Environment Program.

Stern, P. C. (2000). Toward a coherent theory of environmentally significant behavior. *Journal of Social Issues, 56*(3), 407–424.

Viñuales, J. E. (2013). *The rise and fall of sustainable development*. Available at SSRN: ssrn.com/abstract=2200083 or dx.doi.org/10.2139/ssrn.2200083

World Bank. http://data.worldbank.org/

Worldwatch Institute. (2014). http://www.worldwatch.org/agricultural-subsidies-remain-staple-industrial-world-0. Visited 9 Aug 2014.

Chapter 4
User Innovations in Solar Power

Can consumers innovate in complex products? A case study of user innovations in international solar boats competition provides a positive answer. Electric mobility is an emerging global market and includes shipping. It is illustrated how student teams generate know-how on photovoltaic, electric and ship building technologies entailing innovations. The performances measured by the boats' speed during several sailing days increase faster than in the high-tech industries. The teams have created a few dozens electrotechnical and ship construction inventions, and several electrotechnical business have started. The organisation of this competition has generated net social benefit in the region where the events took place. Adoption by the vested ship building industries, however, is slow. The consumer initiatives overcome obstacles caused by high cost and quality imperfections of complex technologies and foster market creation. Consumers are a major source of high-tech innovations.

4.1 Innovating Users

The conventional answer to the question about how firms know about what new items are attractive for customers and how to mix these into saleable products is that it is specialisation of research and development. The research and development department is supposed to investigate these issues and give answers to the firms' managers who then decide about investing in an innovation that eventually satisfies customers' demands. The advantage of companies above individual specialists would be that companies can afford specialists in research and development is argued in the evolutionary theory (Kogut and Zander 1992), as opposed to the view that firms are being organised with the aim to reduce transaction costs (Coase 1937). A different viewpoint on the question above is that the knowledge about the demands and activities is embedded in individuals because this knowledge is generated long before any firm has started research and development activities and it is transformed into companies' know-how through experiences and personal interactions entailing the research and development specialisation. This specialisation can only enhance the process of transformation know-how into products (Polanyi 1966). Companies exploit the tacit consumers know-how for the profit-making activities (Grant 2007).

The argument that know-how is embedded tacitly in personalities has gained popularity in the present knowledge societies. One argument is serendipity. Inventions are often unexpected discoveries. Brilliant personages who invent things by incidents and transform these into products are fascinating stories of creativity and persistence. However, the serendipity cannot explain the broad flow of innovations in the last two centuries because such brilliant people are scarce, and the time lag between the inventions and innovations is generally long, and inventors are rarely innovators (Stobaugh 1988). Measured by number, consumers could have a major stake in novelties because the consumers' knowledge is by far the largest source of know-how, in particular during the last 50 years when many consumers became highly skilled as knowledge workers. Consumers are the major sources of novel receipts for food, health and hygiene and suchlike consumables. The question about how many useful products are invented by the famous personalities, how many by the firms and how many are due to the nameless consumers can be answered in favour of the latter for many utensils (Petrovsky 1994). Numerous inventions for pleasant living and decent housing that have disseminated across societies in the last centuries are due to such tacit tinkerers who are largely neglected in the economic thinking because they have not generated the pecuniary rewards of their inventions as the firms do (Mokyr 2005; Bryson 2010). The consumers could also be the driving force of innovations if their efforts would deliver not solely piece-by-piece works for homes but would also create markets with sufficient demands to generate flows of new products. If this hypothesis is underpinned, the consumers' know-how can be comprehended as an exploitable, cheap and know-how resource for innovations that enables firms to make profit on the markets where consumers have to pay for their efforts. In this train of thought, the positive external effects of the consumer know-how are used for the open source innovations, and these are internalised by the firms. Herewith, possibilities for using the consumers' know-how for the technologically complex products aiming at sustainable development are underpinned.

The innovating consumers were called 'prosumers' (Toffler 1980), but this term lost popularity to the Von Hippel term of 'user innovations'. These terms refer to inventions of individuals who use their and peers' experiences for improvements of products. Such inventions are found in sports, automotive equipment, medical devices, software and so on. The user innovators would not act for profit but for private satisfaction and social motives, such as functionality, trend setting and fun, and they would share their creativity in a network of peers, nowadays called 'community'. Many community members are skilled professionals who are enthusiastic users of products and willing to spend time on the improvements by trial and error (Shah 2000; Luthje et al. 2002; Franke and Shah 2003). About 4–6 % of all consumers in Japan, the United Kingdom and the United States would act as the user innovators. They would invent for the high-value low-volume uses, which are missed by firms because they are considered unprofitable. Frictions with markets would be avoided because the user innovators would innovate primarily for themselves and

4.1 Innovating Users

their peers, whereas firms would innovate for sales on markets. The uptake by the firms outside the community is slow (Henkel and von Hippel 2003; von Hippel 2005; Franke et al. 2006). These observations shed light on those who innovate 'plumber-wise', but they could be myopic about the division of markets if to consider the fast pace of market entry and dissemination of new dishes, fashion, arts, sports etc.

If the user innovations disseminate, such tinkerers could have been a major source of welfare growth throughout the last centuries and could be so in future due to emerging technologies that enable product customization, such as the distributed manufacturing, mobile communication, 3D printers, drones, robotic and so on. If the dissemination of user innovations is negligible because of the division on markets as theorised on the user innovations, not many people are reached and benefited from the advances, and the user innovators' contribution to welfare is low; if the user innovations do disseminate, there is competition between the so-called user and firm innovators. Whether the user innovators are bound to the high-value low-volume community of peers and if their innovations can turn into marketable products, therefore, are important issues for the welfare growth. It is plausible that many consumers commence as voluntary initiatives; some evolve into experts called user innovators with novelties that diffuse in the communities of peers, entailing a number of entrepreneurs who innovate on markets. Fostering this process through experiments aiming to protect the innovators from harsh competition is advocated, and public support of such initiatives is suggested (Brown et al. 2003; Smith and Raven 2012). Subsidies, herewith, are typically public support tools. Awards are typical private support tools; the awarding has grown into a 'prize industry' of USD 1–2 billion a year with prizes for flowers, arts, science, architecture, games, hackers and other communities of peers (McKinsey 2009). Both tools could work in the sense of inducing user innovations. A positive answer to the question whether the user innovators turn into business and the communities of peers generate markets for complex technological products would support the hypothesis that many innovations are due to consumers who become innovators or whose results are internalised by firms. This would underpin the social character of innovating.

The presented case of the user innovations is about an advanced technological hardware that needs large investments: boats powered by photovoltaic technology. The presentation reflects on results of competition between students, professionals and firms in solar boats, ones that use solar power on board with non-grid storage. The solar power boats are considered sustainable innovations that compete with the fuel-based sailing. The competition called the DONG Energy Solar Challenge (DSC) involves hundreds of volunteers and user innovators in international teams. The teams are composed mainly of students from technical and engineering colleges under the supervision of faculties and business professionals. No one of these teams could do research and development. All this tinkering is done by trial and error usually in colleges though professionals and firms are also involved. The competition has gained attention of international media and awards, among these the NUON business 'Innovation Award', the governmental 'Innovative Team Player Energy Transition 2007' and the 'European Solar Prize 2008'. Technical issues related to design and construction of the boats can be found in the Tim Gorter dissertation (Gorter 2014). Herewith, economics is considered. Information about this competition can be found on web: www.dongenergysolarchallenge.nl.

4.2 Markets for Solar Boats

The potential market for the solar boats is sufficient for the large-scale production. It is not for the cargo shipping where powerful combustion engines (500–10,000 kW) for high speed (30–35 km per hour) are used. Although that high speed causes high fuel costs, which generally outweigh the benefits of shorter travel time, clients demand fast shipping services (Krozer et al. 2003). Space on board of ships for the photovoltaic equipment is insufficient to satisfy this power needs though this equipment is sometimes used for appliances on board. The leisure boats can use solar power, as shown in an inventory of 183 solar-powered boats (Gorter 2014). The leisure fleet uses engines generally below 500 kW per vessel, and there is ample scope for designs of photovoltaic equipment on vessels for rowing, paddling, sailing, racing, cruising and luxury yachts. Speed is demanded on leisure markets, and there are solar boats that sail up to 45 km per hour albeit costly, e.g. Czeers in the Netherlands. A few types of the solar-powered boats are shown on Picture Serial 4.1. They vary (Gorter 2014): length 2.13–33 m, width 0.91–22.83 m, maximum draft 0.1–1.2 m, empty weight 98–115,000 kg, full weight 190–185,000 kg, photovoltaic surface 0.6–536 m^2, photovoltaic power 0.05–93.5 kWp, motor power 0.14–162 kW, 1–2 engines, battery capacity 0.076–1750 kWh, cruise speed 3–20 km per hour, maximum speed 3–40 km per hour, person capacity 1–150, price € 2500–24,000,000 (USD 3250–31,000,000).

Inquiry into opinions of yacht owners suggests many benefits of solar power, such as fuel saving, large scope of actions due to unlimited energy resources, durable and low-maintenance electric systems and clean and silent engines (Mallet 2010). The leisure fleet constitutes a large market opportunity for the photovoltaic equipment. The global annual sale of boats is about USD 18.2 (€ 14) billion in 2009. The United States covered about one-third. Boating is a popular way of leisure. Sales have grown by an average of 8 % in 2000s, which is higher than tourism as a whole, but the financial crisis put an end to that high growth rate (Haas 2010; Market data 2014). Assuming that 10 % of the global fleet, i.e. 3.6 million leisure boats, is suitable for the photovoltaic equipment, the costs and performance of the boats with fuel-based engines are compared to the boats with the solar power-based engines. The United States market data on the leisure boats, leisure days and persons (Haas 2010) are scaled up for the global scale. Table 4.1 shows the results based on conservative assumption about the photovoltaic equipment on a vessel because more cost-effective solar power is possible. If the solar power provides auxiliary propulsion, it can substitute nearly one-third of all fuel use, which saves about € 4 (USD 5.2) billion a year and reduces 19 billion tonnes CO_2 emissions a year (nearly equivalent of the CO_2 emission of all global aluminium production). The photovoltaics systems could gain € 28 (USD 36.4) billion market, depreciated in 10 years. Better designs, higher speed and manoeuvrability of sailing are needed.

4.3 Users' Perspective

Research Human transport

Leisure Racing

Picture Serial 4.1 Examples of solar-powered boats. I am grateful to Tim Gorter

Table 4.1 Gasoline or solar power use on the leisure boats; the United States sales are 36 % of the world (*PV* photovoltaic, *bln* billions)

36 mln world fleet scaled up with 13 mln USD	A typical boat		The world fleet			Difference fuel: PV
	Fuel	PV	Units	Fuel	PV	
kW/boat	15	0.8	GW	547	28	(519)
Gasoline kWh/year	2423	792	GWh	86,218	28,179	(58,039)
Gasoline l/year	198	0	mln ton	7054	0	(6)
Annual costs €/year	238	120	€ mln	8465	4270	(4)
CO_2 kg/year	531	0	bln ton CO_2	18,905	0	(19)
Sales photovoltaic equipment € bln			bln €	Nil	28	28

4.3 Users' Perspective

The DONG Energy Solar Challenge (DSC), initiated in 2005 in the Frisian region in the Netherlands, evolved into a biannual international solar boat competition. The Australian solar car competition was the main inspiration. The organisers aimed to link the knowledge and enthusiasm of students and scholars with professionals in the electrical, construction and shipping industries to foster innovations in solar power use, particularly on boats and yachts. The policymakers envisaged economic development in the Frisian region that hosts many ship builders and is a cradle of solar boats worldwide (Fadeeva et al. 2008). The competition covers a 100 m sprint and a 210 km marathon over a period of 5 days. The number of teams participating has grown from 26 in 2006 through to 40 in 2008, to 43 teams in 2010. About two-thirds of the teams are Dutch and one-third from nine other countries. Herewith, the 2012 and 2014 editions are not covered because the route and boat specifications are changed. The number of teams is stabilised. The teams are divided into three classes. Class A has one person on board, class B two people on board. All teams in these classes use the Sun Factory solar panels with Sharp photovoltaic technology offered free of charge. The teams in the class C (top class) used their own equipment. For the 2010 edition, maximum solar power was limited to 1.75 kW in order to foster diversity of innovations on boats instead of introducing more powerful batteries.

Many teams have generated inventions to improve their performance. These inventions, as well as cooperation within the team and external organisations, and the teams' costs are assessed based on 22 team responses to a questionnaire and scaled up for 43 teams that participated in the 2010 edition. Most teams put efforts into propulsion, boat designs and electric system. They have proposed 41 new solutions and have implemented 20 novel products before the 2010 edition, which means that nearly every team had an invention and about half of the teams have implemented these. The solutions addressed mainly new propulsion and electrical systems. The boat design is found most difficult to improve, but a few innovations are introduced, for instance, the hydrophobic solar-powered boats that sail above water surface. The teams spent € 20,000–75,000 (USD 32,500–97,500) per edition. A boat costs € 6,000–100,000 (USD 8000–100,000). An average cost is € 28,800 (USD 37,440) per boat, excluding solar panels in the A and B classes because these are supplied by the sponsor Sun Factory at a cost equivalent of roughly € 2000 (USD 2600) per boat. Most costly is boat construction. These solar boats are more expensive than the marketable boats with fuel engine. Hence, cost reduction is needed to reach the leisure markets. The designs vary from a simple sloop to fancy racing speedboats. When the teams are divided into the university teams with students and staff, business teams and the mixed university–business teams, it is found that the university teams and the mixed teams spent less than half of money spent by the business teams, but their competition ranking is not lower. The rank to expenditure correlation is negative. Students with some supervision are as good tinkerers as professionals. The mixed teams were the winners in three subsequent editions; the business teams won the last two editions.

The performances measured by the maximum and average speed of sprint and marathon are assessed with the official time data. The sprint performance depends a lot on the battery power. The marathon performance, which approaches the leisure uses, depends on the overall boat design. For the sprint and marathon, the average and maximum speed per class and per edition is shown in Table 4.2; time data of some teams were debated in the 2008 edition so only

Table 4.2 Boats' speed in DONG Energy Solar Challenge editions

	Sprint: km per hour			Average change (%)	Marathon			Average change (%)
	2006	2008	2010		2006	2008	2010	
A: maxim.	9	21	14	50	10	11	13	11
A: average	8	11	8	6	5	7	9	31
B: maxim.	11	13	9	−6	10	9	11	6
B: average	7	10	8	6	5	7	9	39
C: maxim.	14	14	16	7	13	16	18	21
C: average	8	14	12	29	6	9	13	45

times of several winning teams are available, and in 2010, the solar power is limited. The maximum marathon speed is 18 km per hour, the average speed 13 km per hour. These approach the usual speed of leisure cruisers. The competition became tougher, and the class C teams became faster due to better designs. The technological change is usually expressed as a percentage change per period or per serial product. The speed has hugely increased due to better designs given the photovoltaic equipment (classes A and B) and due to better equipment with design (class C). Particularly, the marathon speed increased across many teams, not solely the fastest ones.

4.4 Market Perspective

The relevance of this competition for the teams' know-how, innovations and business, as well as for the region, is assessed after the 2010 edition. The relevance for the teams is based on inquiry into opinions of 23 team managers and scaled up for all 43 teams. The managers' opinions indicate that 331 students, 206 academics and 241 businesses were involved. An average team consists of 18 members, of which one-third are professionals. It is a large team compared to the usual research and development teams in high-tech industries. Nearly all teams have generated new know-how, two-thirds have developed technologies, more than half of the teams shared knowledge with others, and one-third of the teams have plans to start a business. The results suggest that a dozen of firms could be created due to this event; a few participating teams have already started a business. The businesses in the university teams and mixed teams are not solely sponsors but usually also suppliers of equipment and engineering services for construction. Branding of the firm's name is their main interest. Several firms have generated know-how, but many are cautious about profitability of their participation though all except one have confirmed their participation in the next editions.

The regional benefits outweighed the costs. The costs of organisation per event are nearly one million euros. The main benefits are the teams' investments in the boat technologies, teams' payments for participation in the events and branding of the sponsor names. The latter is assessed with the lowest value advertisement prices. These benefits outweighed the costs by three times. Both, the private sponsors and public financiers, have gained net benefits. Unaccounted are the indirect benefits: the follow-up revenues of the partnering sponsors and financiers, as well as the know-how and innovations of participating companies. These can be large, for example, a company mentioned that a few million euros order option is gained due to networking during the event. The growing number of teams, more participants per team and an increasing number of volunteers indicate high public interests in technologies if innovations are tangible and perform. The dissemination of the solar power across the vested boats and yachts businesses, however, remained slow despite efforts put in organising a working group of yacht builders and photovoltaic equipment suppliers (among them international prize winners) with support of technical universities, business workshops with boat and yacht constructors and policymakers and presentations on business fairs. Whether these events have had effects on multiplication of solar power services for boats and yachts cannot be claimed, but it might have promoted such services. Meanwhile, a market of the photovoltaic equipment services on boats and yachts has emerged.

The possibilities of exporting the events are discussed with team managers of all countries. The United States-based 'Solar Splash' is the only alternative, so far. Incidental events took place, in Zeeland province in the Netherlands, Brasil, Italy and Poland. The exporting of such organisation is found laborious. The barriers are not primarily the organisation and funding of an event because these can be found on many places, but large social involvement of volunteers and broad stakeholders' participation are needed to maintain activities. These are not easy to get. The social embedment is essential because dependency on a few senior persons, large businesses and authorities runs the risk of collapse when interests of a few stakeholders change. It is important to overcome such changes. When this initiative emerged, it received warm support from the photovoltaic equipment producers, from an electricity producer and from universities and regional authorities, but the boats and yacht producers have shown a 'wait and see' attitude. The constellation changed after the first 2006 edition when many regional boat and yacht firms were attracted, but the electricity producer pulled out, whereas another electricity producer and a bank partnered the organisation. After the even more successful edition in 2008, fortunes of the event have changed again. The electricity producer and bank withdrew their support because of reorganisations, and the authorities piled up verifications, which nearly led to the demise of the organisation. The continuity was ensured by the regional boats and photovoltaic equipment suppliers and a few regional partners. All of them were small- and medium-scale innovative companies and nongovernmental organisations, not the established businesses in the Netherlands. Foreign partners were more supportive, such as DONG Energy (electricity producer from Denmark), the European JRC Energy Institute, the Jordanian Royal Court, a regional authority in Liyang, China and others. Hence, the organisers put effort into broadening stakeholders' basis through services, such as manual and classes for self-made solar boats, research on solar boat designs, support of similar events abroad, promoting solar-based start-up and so on.

4.5 Conclusion

It is discussed whether consumers can create markets for sustainable innovations. An observation is that consumers, being users of products and services, exchange know-how about products in use entailing numerous performance improvements. These tacit improvements labelled as the user innovations are encountered in food, leisure, sports, software and so on, but if the improvements are not disseminated, their positive effects on welfare effects are limited. The question whether the user innovators can create markets of complex technological products is largely positively answered under the condition that the efforts are substantially large and continuous.

This answer is based on the competition with solar-powered boats that covers sprint and a 210 km marathon during 5 days without the use of grid. The expectation that the participating students and professionals create an innovation spur is confirmed. The technological change measured by speed is very high even by the high-tech standards. Several dozen inventions related to photovoltaic systems and boat construction are developed and implemented, a few hundred businesses got involved, a dozen business start-ups are pursued, and monetary benefits of this event outweighed the costs by three times. High involvement of students and much public appreciation indicate that such events also generate educational spin-off entailing capacity building for innovative spur in the future. There are two main impediments. Firstly, copying of such event is difficult. Support of volunteers and stakeholders is prerequisites for continuity of such communities of peers. Secondly, an issue is that the newcomers and the small firms in the yachts and ship building have embraced the solar power designs but the vested shipping companies are slow adopters. The case underpins viability of the hypothesis that the consumers' inventions are the major source of innovations. It illustrates that the citizens' initiatives if supported by stakeholders enable to overcome obstacles caused by high cost and quality imperfections of complex technologies and foster creation of markets of sustainable innovations, but the diffusion of innovations to the vested firms and interests takes much time.

References

Brown, H. S., Vergragt, P., Green, K., & Berchicci, L. (2003). Learning for sustainability transition through bounded socio-technical experiments in personal mobility. *Technology Analysis and Strategic Management, 15*(3), 291–315.

Bryson, B. (2010). *At home: Short history of domestic life* (1st ed.). New York: Anchor Books.

Coase, R. H. (1937). On the nature of the firm. *Economica, 4*(16), 386–405.

Fadeeva, O., Brezet, J. C., & Krozer, Y. (2008). Challenging as a way to invoke innovations. In R. Sullivan (Ed.), *Between the market and the state: Corporate responses to climate change* (pp. 75–85). Sheffield: Greenleaf Publishers.

Franke, N., & Shah, S. (2003). How communities support innovative activities: An exploration of assistance and sharing among end-users. *Research Policy, 32*, 157–178.

Franke, N., von Hippel, E., & Schreier, M. (2006). Finding commercially attractive user innovations: A test of lead user theory. *Journal of Product Innovation Management, 23*, 301–315.

Gorter, T. (2014). *Performance evaluation of photovoltaic bots in an early design phase.* Dissertation, University Twente, Enschede.

Grant, K. A. (2007). Tacit knowledge revisited – We can still learn from Polanyi. *The Electronic Journal of Knowledge Management, 5*(2), 173–180. Online at www.ejkm.com

Haas, G. E. (2010). *A snapshot of recreational boating in America –Background.* Colorado State University, mimeo.

Henkel, J., & von Hippel, E. (2003). *Welfare implications of user innovation.* University of Munich and CEPR London, M.I.T. Sloan School of Management, mimeo.

Kogut, B., & Zander, U. (1992). Knowledge of the firm, combinative capabilities, and replication of technology. *Organization Science, 3*(3), 383–397.

Krozer, Y., Maas, K., & Kothuis, B. (2003). Demonstration of environmentally sound and cost-effective shipping. *Journal of Cleaner Production, 11*, 767–777.

Luthje, C., Herstatt, C., & von Hippel, E. (2002). *The dominant role of "local" information in user innovation: the case of mountain biking* (MIT Sloan School of Management, Working Paper, 4377-02). Cambridge, MA.

Mallet, V. (2010, September 22, Wednesday). Demand for more sustainable boats and equipment grows. *The Financial Times*, p. 2.

Market data. Visited 25 Jul 2014. Global recreation boats www.bharatbook.com; global yachts market: http://www.hugohein.com/partners/yacht.market.specs.htm; yacht in: http://www.nautica.it/superyacht/509/mercato/analysis.htm; the US sales in: Boat Building Industry Snapshot, June 2010, Western Washington University. About diesel http://www.americanboating.org/fueltax.asp

McKinsey. (2009). *And the Winner is... capturing the promise of philanthropic prizes.* Sydney: McKinsey, mimeo.

Mokyr, J. ((2002) 2005). *The gifts of Athena* (5th ed., pp. 1–27). Princeton: Princeton University Press.

Petrovsky, H. (1994). *The evolution of useful things* (1st ed.). New York: Vintage Books, Random House.

Polanyi, M. ((1966) 1999). *The tacit dimension* (2009 ed.). Chicago: The University of Chicago Press.

Shah, S. (2000). *Sources and patterns of innovation in a consumer products field: Innovations in sporting equipment* (Massachusetts Institute of Technology Sloan Working Paper #4105). Cambridge, MA.

Smith, A., & Raven, R. (2012). What is protective space? Reconsidering niches in transition to sustainability. *Research Policy, 41*, 1025–1036.

Stobaugh, R. (1988). *Innovation and competition* (1st ed.). Boston: Harvard Business School Press.

Toffler, A. (1980). *The third wave* (1st ed.). New York: Bantam Books.

von Hippel, E. (2005). *Democratisation innovations* (1st ed.). Cambridge, MA: MIT Press.

Chapter 5
Tacit Inventors in Regions

How can regions use tacit knowledge for innovations? The regional know-how in the stakeholder networks and in consortia for innovative clusters is discussed based on sustainable innovations for the tourism mobility: transport means, arrangements and communication technologies. The stakeholder networks generated 73 sustainable innovations; out of these, 38 (52 %) were successful measured by 10-year activity after the start. Social organisations and small firms are successful innovators albeit the scale of operations is small. The network policy is cost-effective and affordable in all regions. The consortium development followed the EU policy on innovative clusters. This has generated 15 consortia with 65 small- and medium-size enterprises and many experts. Out of these, 6 consortia were arranged and 2 were acting after 10 years. The success rate of inventors has dropped, and this policy is more costly, but the operations are larger than in network policy, and the successful consortia covered costs of this policy. Sustainable innovations for the regional development are generated when local tacit know-how is fostered. Scouting inventors, developing skills and support of the knowledge spillovers are promising tools for the regional economic development.

5.1 Clusters and Networks

The regional economics is highly politicised. When the traditional businesses in a region decay and capable people move out, the regional politicians are blamed. In Belgium, for example, controversies between the previously deprived Flanders that became wealthier than the formerly rich industrial Wallonia undermine the national state (Aernoudt 2008). Similar controversies are observed in Italy, Spain and the United Kingdom. In Europe, many traditional industries on the territory between London and Milan collapsed during the last 50 years. Innovation hubs emerged elsewhere, and the regional disparities in income and capabilities enlarged in Europe (Marelli 2004). Creating regional innovation hubs is often advocated as the method for economic development, but the successful hubs are difficult to predict let alone to plan them. For instance, hardly anybody could foresee the successful communication hub in Tampere (Finland) in the 1970s, the successful but diverging specialisations of the oil and gas hubs in the neighbouring Stavanger (Norway) and Aberdeen (Scotland) in the 1980s (Hatakenaka et al. 2006) or the wind energy hub in Navarre (Spain) instead of in the neighbouring Catalonia with the similar natural resources and huge business academic centres in the 1990s (Faulin et al. 2006).

Much is theorised about turning around the economical periphery regions into high-income communities (reviewed in Malecki 1991; Scott 1998). Two basic approaches with many variants can be found. The conventional approach is specialisations. Districts of bakers, tailors, bankers and suchlike have been created throughout centuries because the spatial concentration enables firms to attract labour, access suppliers, gain scale and so on. These agglomeration advantages would explain the attraction of dense city centres despite congestion, crime, pollution and other external effects. The assumption that the agglomeration advantages also apply for know-how is the foundation for the growth poles theory in the 1960s, which is applied in the cluster policies. Infrastructure is funded from public resources aiming to attract know-how for companies, for example, Dassault-based aerospace cluster in Toulouse (France), and, vice versa, research centres outside cities are created to attract business, such as the Joint Research Centre near Milan (Italy). Comparison of the regional strength and weaknesses, opportunities and tensions to underpin the regional specialisation has gained popularity in policies (Porter 2000). Europe has championed the cluster policy, less so Japan that is focused on the cooperation between business and authorities and even less the United States that prioritises the research and development (Rosenfeld 2002; EU 2003; Mitsui 2003). Many billions of euros are spent in Europe on pulling firms and scientists to the locations with an adjective of 'high-tech' for a valley, park, campus or suchlike (Laffitte 2006). Although this policy suggests the regions become more specialised, similar technologies are usually considered, such as informatics, nanotechnology, life science, and mechatronics. The cluster policy brings rewards to a few regions that generate much national funding, but the overall results are disappointing. Studies that relate the number of patents and innovations to employment and business turnover show positive impacts of the cluster policies on the vested businesses in a region, but hardly positive impacts across businesses and regions (Baptista and Swann 1998; Moreno et al. 2005). It implies that the cluster policy strengthens the vested firms in the regions, ones that get extra funding, but impedes ones in the other regions. Case studies do not show positive effects of the clustering on the regional innovations (Malmberg and Maskell 2002; Martin and Sunley 2003); they also indicate too little spatial dependencies between firms to justify the cluster policies aiming at innovations (Niosi and Zhegu 2005; Bekele and Jackson 2006), and studies underpin that firms and experts are mobile and choose locations efficiently even if large public funding for resettlement is available (Martin et al. 2008; Webb 2008). Large public funding of the regional innovation clusters is insufficient to generate innovations and can be wasteful.

Another approach advocates the development of skills tacitly available in the region and attracting new skills from outside the region with the aim of creating innovative networks (Lawson and Lorenz 1998). This networking, however, is risky. It can fail despite policy support and expected advantages of cooperations because economic impediments are aggravated by social barriers, such as perception of an unequal partnership, frustration of leaders about laggards, miscommunication about implementation and others (Georg et al. 1992). The successful innovations would be due to interactions between the regional businesses, policymakers and experts, a notion that is popularised as a 'triple helix', because such interactions would generate knowledge spillovers entailing innovations. Generating the regional know-how through education, schooling, research and suchlike knowledge-based activities aiming at larger skills is advocated. Development of this know-how and fostering knowledge spillovers would be the tasks of the regional economic policies (Cooke and Morgan 2002; Hospers 2004). It is argued that large creativity and research projects would attract creative professionals whose networks generate regional development (Florida 2005), that the trustful policies would foster such interactions across specialisations and enable risk-taking investments (Babcock-Kumish 2006) and that cooperative arrangements on market niches would foster organisations to adopt novelties and take risky investments in sustainable innovations (Kemp et al. 1988; Jorna and Faber 2006). These sympathetic notions could be wishful thinking when measurable successes are rare. Unfortunately, this is the case.

5.2 Policy on Tourism

The presented case is about sustainable innovations in tourism. The innovative, sustainable tourism is encouraged through the network policy followed by the cluster policy in the Frisian region in the Netherlands. Being one of the lowest spenders on research and development in the northern Europe with high net youth emigration and unemployment, it is a typical periphery region. Opportunities are expected related to the know-how in agriculture, transportation and tourism, for instance, (1) foods and meals based on regional crops and animals, (2) wildlife for the medical and personal care, (3) tourism with sport equipment and transport means. (4) the distributed smart grid services instead of large-scale infrastructure, (5) eco-efficiency management in distribution, (6) communication on distance for education and care, (7) localised business centres, (8) customised transport for less mobile people and (9) arts and crafts related to space and nature (Krozer and Tijsma 2005). These opportunities are approved in the regional policy 2001–2006 with focus on the agro-food, water technology, renewable energy and tourism mobility. Herewith, the tourism mobility is addressed, the water and energy cluster policies are discussed elsewhere (Krozer 2010) and the agro-food policy did not take off.

The tourism business with many small- and medium-size enterprises is the largest private employer in this region. It attracts about two million tourists a year and adds about €1700 (USD 2200) income per inhabitant, equivalent to 6 % of the total regional income. The regional policy has addressed all kinds of tourism, not solely the ecotourism and sustainable tourism defined as being specialisation aiming at the touristic experiences of nature and as tourism with respect to environmental qualities, respectively (e.g. Hardy and Beeton 2001; Weaver 2005).

The policy is focused on water tourism due to many waterways and sailing tradition in this region. The positive economic effects of tourism on economic development are not obvious. Many positive effects are assumed, such as backward linkages to the local economy, adaptivity to local conditions, funding biodiversity protection and so on, but there are social costs of infrastructure and equipment, social tensions when tourists interfere with local culture and environmental impacts. The balance of social costs and benefits depends on many factors, such as the scale of investments to attract tourists, leakages of expenditures outside the region, the value of tourist services to the facility costs, competition of tourism with other sectors and communities' and stakeholders' involvement. Scholars have developed frameworks for assessing and balancing these costs and benefits with the aim to underpin the positive effects of tourism development (Telfer and Wall 1996; Hunter 1997; Gnoth and Anwar 2000; Turnock 2002). The overall positive results, however, are disputed because productivity of tourism is generally low compared to other activities, whereas the high public expenditures in tourism crowd out other possibly productive public and private investments (Lanza et al. 2003; Dwyer et al. 2004). It is also argued that the escalation of tourist services impedes distribution of wealth because income is increasingly concentrated in the hands of large corporations and transferred outside the tourist resorts (Baidal 2003; Papatheodorou 2004). The qualities of resorts also decay after some time, which is difficult to prevent, but restoration is difficult because these qualities are vulnerable to pressures of the intensive tourism. To prevent the decay, innovating for the tourism qualities is advocated, but these involve high costs that are often unbearable for the individual entrepreneur in tourism (Agarwal 2002; Kruger 2005).

With regard to these experiences, the Frisian regional policy on tourism aimed to generate income, but avoid social tensions and environmental degradation. The traditional focus on infrastructure for the development of the water tourism such as dredging sludge to deepen canals, constructing marinas and aqueducts that separate water and road traffic is shifted towards more quality in tourism services with the use of high-tech products. Development of new tourism service is addressed, which is usually less costly than technology development. The regional inventors pursuing sustainable innovations are supported. The regional inventors are considered all individuals who submit proposals for a novel product and service, excluding research and education proposals though these can generate inventions after some time. Two types of policies are introduced. During 2001–2003, the policymakers were made responsible for attracting inventors and could make small grants available to inventors for sound proposals. The inventors could use these grants as seed money for starting activities. This is referred to as the network policy. This one aimed to generate know-how that is tacitly available in the region. The subsequent cluster policy during 2003–2006 aimed at innovative clusters of experts and businesses. This one pursued the European Union procedures for grant applications and assessment, e.g. the applications were assessed by a commission of experts and policymakers. Final decisions are made by the regional politicians. The policy success can be indicated by the number of granted inventions compared to applications (input criterion) and by the granted inventions compared to innovations in operation (output criterion). The latter are relevant for economic development. Hence, the success is measured by the inventions that operate in 2014 compared to the granted ones, which is about 10 years after the grant.

5.3 Inventors in the Network Policy

In the network policy, persons and organisations pursuing valuable, commercially unavailable products are assumed to be potentially successful innovators. The policymakers supported the promising inventors with contacts to stakeholders and grants for seed money. To avoid collisions between the supporting and regulating roles of policymakers, these functions are allocated to different policy departments. A checklist without formalities is used for the grants. The criteria were value added for the region, innovation, distribution of wealth through participation of the regional small- and medium-size enterprises and environmental qualities. Out of the total €20 (USD 26) million annual regional policy support of tourism, about 10 % was for activities pursuing quality, and within this budget, about 10 % is used for the seed money. About half million euros during 3 years are spent on such grants as co-funding of the private and public investments. The granting evolved in bits and pieces when policymakers found that the proposal is attractive for an invention. The grants were small, some were below €5000 (USD 6500), and they were rarely above €20,000 (USD 26,000) per inventor. Many inventors found these grants relevant for the start because it was a sign of confidence used to attract private funding and involves users for quality testing, tune inventions to customer demands and suchlike. The inventors were found among the non-governmental organisations, tourist organisations, firms, educational and research institutions, individual experts and policymakers in their civic functions. Fraud was not found probably due to transparency of the process.

The granted inventions are divided into three specialisations: the transport means, arrangements and communication tools. The main results are summarised in Table 5.1. Appendix lists all inventions: the subject, the topic of invention, the inventor, is it a starter or active for some time, leader in operation and expected barrier in execution, based on Krozer et al. (2005).

A total of 73 inventions were granted. There are 25 grants for novel transport means and 10 of them operate, after 10 years 32 for arrangements and 21 still operate and 16 granted inventions for the communication tools and 7 still operate. In total, 38 granted ideas operate after 10 years. The success rate is 52 % of all

Table 5.1 Granted inventions in tourism

Project	Start (new one)	Active (from the past)	Total
Biking	2	3	5
Shipping	9	6	15
Vehicles	5		5
All transport means	16	9	25
Care	3		3
Culture	13	10	23
Nature	3	3	6
All arrangements	19	13	32
Physical	1	1	2
Virtual	14		14
All communication	15	1	16
Total	50	23	73

inventions, which is exceptionally a high rate compared to financing of sustainable innovations as discussed in Chap. 9. About three-quarters of the innovations were new ideas (start), and one-quarter of all were ongoing for some time (active). Various stakeholders made proposals and received a grant: nine non-governmental organisations, seven policymakers, six companies, six tourist organisations and five experts. The social organisation and firms were particularly successful, the experts and policymakers rather unsuccessful. The transport means cover biking, shipping and vehicles focused on the electric engines and biofuel. The flat Frisian landscape with many lakes is suitable for the bikes and ships. Most successful projects are related to shipping, aiming at solar power on sailing boats, electric ferries and cruises. The non-governmental organisations were granted seven inventions and many were successful, the experts' five inventions, but these were rarely successful and the shipping firms three inventions and all were successful. The arrangements cover inventions for care such as the use of a spa and revalidation, culture tours such as religious or monument tours and nature-oriented arrangement. Most arrangements have addressed the cultural points of interest. Most granted proposals were done by the non-governmental and tourism organisations with eight and seven successful inventions, respectively. The policymakers' proposals were rarely successful. The communication projects are focused on using the information and communication technologies mainly for guiding tourists and as a marketing tool. The tourism organisations received most grants, but only three out of eight are successful. The success rate of experts and policymakers was particular low.

There are barriers for the successful innovations. The user requirements were considered by the inventors as being the main thresholds, in particular seasonal fluctuation in tourism and poor product quality as perceived by the users. The development cost is perceived the major barrier by companies. A few inventors criticised strict regulations, but others argued that authorities should be more demanding.

The policymakers and experts' inventions often failed because of poor entrepreneurial capabilities. It should be noted that the assessors did not applaud many inventions because the lack of success could reflect on their professional performance. The high success rate of the arrangements, therefore, could be due to the strict screening, but good proposals could fail because of the professional biases, e.g. rejecting proposals when they did not follow the standards taught on courses or in policymaking.

The scale of operations and profitability are unknown, but most sustainable innovations are small scale. It could be that each innovative activity considered separately is not profitable and would be assessed unattractive when screened against formal assessment criteria, but many survived because they are compounded into a service package. Compounding the small-scale activities creates attractive tourism propositions in the region that lacks large-scale attractions. The answer to the question about how many granted inventions would fail without the grant is speculative, but a motivated guess is that many starting activities would not take off or fail without this support, which is 5 out of 16 in the transport means, 10 out of 19 in the arrangements and 14 out of 15 in the communication tools. The virtual communication tools would probably be missed without the grants though the grants for this specialisation were small and less successful compared to the others. The network policy attracted many tacit inventors and enabled to use their know-how on the cultural and environmental qualities for income generation. Such policy is affordable to all regions. Assessing profitability of individual inventions is a deficient approach because it disapproves compounding of individual services into service packages that benefit region as a whole. Better is assessing the social costs and benefits of compounded assortments.

5.4 Consortia in the Cluster Policy

The innovative cluster policy in this region started for an opportunistic reason that the regional government could get a European Union co-funding in 2003. The cluster policy in tourism ended 3 years later, it is still ongoing in water and energy and it started on agro-food and on aging. Instead of awarding the inventors with seed money, this policy aimed to create business clusters through grants for consortia with several businesses and experts. The grants were usually 20 times larger than the seed money in the network policy. In line with the European granting procedures, applications were assessed by commissions of policymakers and experts with recommendations to the regional authority about granting. In the assessments, the value chain approach is adopted (after Porter 1992), which meant that the users had to participate in the project and present their benefits next to the social and environmental benefits. The partner search for the consortia was facilitated by external consultants. The author, one of the consultants, made 15 attempts to arrange consortia; all are based on initiatives of an entrepreneur. The consortia are briefly described to illustrate the diversity arrangements and possible sustainable innovations in tourism.

Nine attempts to set up a consortium were stopped without application. 'Care and tourism' aimed to develop vacation centres for the handicapped and patients after hospital treatment. The initiative was taken by two entrepreneurs who proposed to establish a revalidation centre in a tourist area. They agreed on a consortium but found this business too risky. 'Chinese tourism' aimed at leisure tours for Chinese entrepreneurs who attend international fairs in the Netherlands with the focus on waterworks and agriculture. The initiative is taken by a tour operator. Market study has shown interest of the Chinese businessmen, and a 1-day tour for a dozen Chinese journalists is arranged with publicity in China. Seven organisations have shown interest to participate, but none agreed on taking the lead. 'Farms for care' aimed to combine work and education for people of limited mental abilities and this way to shorten the list of those waiting for facilities and creating extra income for farmers. The initiative is taken by an agricultural lobbyist and an entrepreneur. After contact with farmers, social workers and teachers, two organisations were found interested, but no one took leadership. 'Lotus for Fryslân' aimed to develop aquaculture of lotus plants on the agrarian waterways mainly for the cleanup of waterways. The initiative came from a group of farmers. Contacts were established with Chinese experts who suggested lotus varieties that would be suitable for the Dutch climates. The entrepreneurs were interested, but no one was willing to take the lead. 'Marshmallow' aimed to cultivate marshmallow as a natural sweetener and flavourant. The agrarian entrepreneurs mentioned above initiated this project with the aim to introduce regional foods from brackish environment. Market research into the marshmallow production and sales revealed possibilities. A pilot was suggested, but no entrepreneur wanted to take the lead. 'Networked society' aimed to use the available broadband for services in rural areas, e.g. shopping, learning, mobility and so on. The initiative was taken by a telecom company and an expert. After 2 years of discussion, five business and expert organisations have reached agreement on a pilot, but the pilot is delayed because no business wanted to take the lead. 'Street event' aimed to use public space for culture and sports events, which would strengthen social cohesion and foster local development. The initiative is taken by a social worker and an expert. Several sports and cultural organisations that are approached were interested, but a business leader is not found. Such initiatives emerge from time to time with little success. 'Village transport' aimed to provide cheap short-distance travels for elderly people with electric vehicles. The initiative was taken by the management of an elderly home. It would act as the leading entrepreneur. Several entrepreneurs in the local mobility services agreed to take part in a pilot, but the preparation was stopped when the management is laid off because a few elderly houses merged.

Several consortia are organised, but they failed to reach continuity. 'Frisian Aqua Fair' aimed to establish an international cooperation of policymakers and companies aiming at management of water bodies on islands for tourism. The idea that

5.4 Consortia in the Cluster Policy   67

originated at the UN University is taken up by the regional water business cluster consisting of five businesses and a few expert organisations. A grant enabled an international meeting, but developing an international consortium that relates water and tourism was too laborious. This project has generated a spin-off project on islands in Europe. 'Electric sailing' aimed to introduce electric boats based on hydrogen. The initiative is taken by a few entrepreneurs and an education institution. The hydrogen sloop was granted, and the boat was launched during the DONG Energy Solar Challenge, but technical problems constrained follow-up. 'Solar power sailing' aimed to apply solar energy for stand-alone electricity on yachts. The initiative was taken by an entrepreneur. Prototype of a solar energy yacht is constructed, and cooperation with a sailing school is arranged. A pilot project is granted and is successfully executed, but scaling up failed because of financial constraints and illness of the initiator. 'Village centres' was purposed to restore historical buildings for cultural, educational and other public uses. The initiative is taken by an entrepreneur aiming to foster quality of life in villages. After contacts with schools, care centres and the organisation of villagers, a few entrepreneurs were interested, but the business was unfeasible because the diluted rural population has a low income. A few years later, another consortium is created. 'Virtual tourist office' aimed to establish a virtual tourist information system on the Wadden Sea, the UNESCO World Nature Heritage that covers Denmark, Germany, and the Netherlands. The lead is taken by an entrepreneur, and 14 entrepreneurs agreed on a pilot. The grant was approved, but a competitor sued the grantee. This opposition was unsuccessful, but the consortium dissolved.

Two consortia attained continuity. 'Wadventures' aimed at boat tours on the Wadden Sea for pupils. The tours use information and communication technologies for learning about nature. The initiative was taken by an entrepreneur and expert. A consortium of six entrepreneurs and educational institutions is created. After a laborious granting negotiation, pilot tours are successfully executed. 'Wetter bus' aimed to bring tourists along the natural and cultural attractions on antique busses, thus to reduce car travels. The initiative was taken by a tour operator who got support of his suppliers. A consortium of four organisations in the supply chain of this tour operator received a grant. The pilot was successfully executed and scaled up. The cluster ideas, number of enterprises and results are presented in Table 5.2.

The cluster policy illustrates that there is much entrepreneurial interest in the regional sustainable innovations as 65 small- and medium-size enterprises have participated in the consortia, as well as many experts and educators. However, it is difficult to reach agreements about grant application between partners and even more difficult to continue after the granting. Out of 15 consortium proposals, 6 received a grant, and out of these 2 still operate, which is 13 % of all attempts. This percentage is in line with statistics on percentage business start-ups in the total of innovating firms as shown in Chap. 9. There are also 2 spin-offs. This policy is also affordable to many regions because the transaction costs were about €0.3 million and €0.2–€0.3 million per project is granted, which covered 50 % of the project

Table 5.2 Cluster ideas for sustainable innovations

Names	Description	SME	Result and barrier
Care and tourism	Revalidation in vacation housing	2	Stop: technical
Farms for care	The handicapped work and learn	2	Stop: no leader
Frisian tours	Tours for Chinese businessmen	7	Stop: no leader
Lotus cultivation	Aquaculture for wastewater	2	Stop: no leader
Marshmallow	Aquaculture for flavours	3	Stop: no leader
Networked society	Broadband for virtual service	5	Stop: no leader
Street events	Areas for culture and sports	2	Stop: no leader
Village centres +	Multifunctional rural buildings	2	Stop: technical
Village transport	Elderly people local transport	4	Stop: technical
Electric sailing	Hydrogen electric boats	5	Stop: technical
Frisian Aqua Fair[a]	Management of water bodies	5	Stop: technical
Solar power sailing[b]	Stand-alone electric yachts	2	Stop: technical
Virtual tourist office[b]	Internet-based marketing	14	Stop: technical
Wadventures[b]	High-tech tours on sea	6	On
Wetter bus	Bus tours for nature and culture	4	On
Total		65	

Note: Words in bold means granted
[a]Spin-off
[b]Initiated by the network policy

budget. The two successful consortia did not evolve into a business cluster, but their revenues cover the costs of grants. In the tourism cluster policy, only one more consortium called 'Rural games' is granted, but it failed. The grants for tourism stopped because of other priorities. The barriers for consortium development are the unexpected technical and economic problems in seven cases and lacking entrepreneurial leadership in six cases. Small or big consortia hardly matter for the success rate. The two successful consortia had average five entrepreneurs per consortium compared to average four entrepreneurs for all attempts. The granting process, however, is biased towards the large consortia. The grants are given to the consortia with an average of more than 7 firms per consortium although the large consortia are often unsuccessful. Finding the entrepreneur who is willing to lead the consortium is difficult. The search for the entrepreneurs through institutions, e.g. the Chamber of Commerce, has generated many names but little success. The snowball search through personal contacts runs the risk of trapping into a network. This search is easier when events promote sustainable innovations. It is also observed that the policymakers and experts tend to push their own ideas although these ideas are rather unsuccessful and impede arrangements in consortia. Another barrier is the time-consuming grant approval process.

5.5 Conclusions

The question about how to generate tacit knowledge for regional development is discussed. The case is focused on sustainable innovations in tourism, which cover high-tech products and services for tourist transport, arrangements and information and communication. The success rate of the stakeholders' networks and consortia of firms and experts is measured by the number of granted inventions that are active after 10 years. The networks were supported by the regional policymakers with seed money and contacts without formalities. This support was exceptionally successful: 52 % of 73 granted inventions operate. The most successful inventors were the non-governmental, tourism organisations and firms; the least successful ones were experts and policymakers. Many successful nonbusiness inventors indicate much tacit knowledge in the region, which can be used for sustainable innovations. Many inventions are small and low profit, but they have large contribution to tourism when considered together as an assortment. This policy is affordable in all regions because costs are low. The positive experiences with the network policy confirm the Law of Parkinson that 'success breeds bureaucrats'. The subsequent cluster policy followed the European Union procedures: consortia of businesses and experts organised by consultants aimed to submit grant applications that were assessed by experts and policymakers. Based on the development of 15 consortia with 64 firms and many experts, it is experienced that the success rate dropped to 13 % despite more efforts put into the organisation of consortia. The costs per project in the cluster policy were 1–20 times higher than in the network policy, but the income per project was also larger. The case illustrates that various sustainable innovations in tourism are possible at low costs, based on the regional tacit knowledge if policies foster inventors through arranging of the knowledge spillovers. Many firms are inventive, and even more inventors can be found in social organisations if the seed money is granted without formalities and policymakers provide assistance instead of control. This approach generates more successful results, and it is more cost-effective than organising consortia with formal grant applications because entrepreneurs are reluctant to take responsibility for consortia which are considered risky. Scouting inventors, developing entrepreneurial capabilities and transforming these capabilities into sustainable innovations generate economic development in periphery regions.

Appendix

Granted project ideas for sustainable innovations in tourism during 2001–2003; those in bold are still active. The non-governmental organisations are indicated as 'NGO' in the table below.

Table 1 Sustainable innovations in tourism transport

Transport	Project	Inventor	Use of local quality	Status at granting	Stakeholders	Barrier	What barrier
Bikes	Rickshaw in Leeuwarden	Transport company	No special	Active	Companies	Users	Season peak
Bikes	Fun bikes Mopark	Tourist companies	Lakes	Start	Companies	Users	Product quality
Bikes	People mover Terschelling	NGO	Carless island	Active	Companies	Users	Season peak
Bikes	Solar people mover Alde Feanen	Experts	Landscape	Start	Users	Companies	No entrepreneur
Bikes	Ameland mitka	Experts	Carless island	Active/stop	Companies	Users	Product quality
Bikes	Stand-alone toilets NOLIMP	NGO	Lakes	Active	Authorities	Companies	Extra cost
Ship	Electroboats Alde Feanen	Tourist companies	Lakes	Active	Companies	Authorities	Spatial policy
Ship	Solar boat Suwald	Shipping company	Canals	Active	Users	Users	Season peak
Ship	Solar boat Mopark	NGO	Canals	Active	Users	Users	Season peak
Ship	Solar boat Fryske Gea	NGO	Lakes	Active	Users	Authorities	Spatial policy
Ship	Sustainable marina	NGO	Lakes	Start	Authorities	Companies	Extra cost
Ship	Ballast water sites	Tourist organisation	Lakes	Start	Authorities	Companies	Extra cost
Ship	Brown fleet harbour	Authority	Lakes	Active	Authorities	Users	Extra cost
Ship	Solar cell marina	Authority	Lakes	Start	Authorities	Users	Extra cost
Ship	Noah arch	NGO	Lakes	Start	Authorities	Users	Uncertain sales
Ship	Solar boat challenge	NGO	Skate tour	Start	Companies	Users	Marketing
Ship	Digit water map	Tourist organisation	Lakes	Start	Companies	Users	Season peak
Ship	Solar sailing Torenvalk	Shipping company	Lakes	Start	Companies	Users	Product quality

Appendix

Ship	Funboats Alde Feanen	Experts	Lakes	Start	Developers	Companies	Uncertain sales
Ship	Solar boat Princenhof	Shipping company	Lakes	Start/stop	Companies	Authorities	EU policy
Vehicles	Rapeseed fuel bus	Authority	No special	Start	Authorities	Companies	Extra cost
Vehicles	Electrocars Vlieland	Transport company	Carless island	Start	Companies	Users	No entrepreneur
Vehicles	Rickshaw Alde Feanen	Transport company	No special	Start	Companies	Users	No entrepreneur
Vehicles	Solar ice man	Experts	No special	Start	Developers	Companies	No entrepreneur
Vehicles	Transfer to Ameland	Experts	No special	Start	Developers	Companies	No entrepreneur

Table 2 Sustainable innovations in tourism arrangements

Arrangements	Project	Inventor	Quality	Status at granting	Stakeholders	Barrier	Limitation
Care	Blind people attraction	Authorities	No special	Start	Authorities	Users	No entrepreneur
Care	Handicapped attractions	Authorities	No special	Start	Authorities	Users	No entrepreneur
Care	Spa on Ameland	Tourist organisation	Nature	Start	Users	Companies	No entrepreneur
Culture	Historic harbours	Experts	Dispersed sites	Start	Authorities	Users	Extra cost
Culture	Hanse passage	Authorities	History	Start	Authorities	Users	Uncertain sales
Culture	Theatre 11 – cities	NGO	Skating tour	Start	Authorities	Users	Uncertain sales
Culture	Wooden constructions	Authorities	History	Start	Authorities	Users	Extra cost
Culture	Culture tour in busses	Tourist companies	Dispersed sites	Start	Companies	Users	Uncertain sales
Culture	Steam tours	Tourist companies	No special	Start	Companies	Users	Extra cost
Culture	Classic shipyard Terherne	Tourist companies	No special	Start	Companies	Users	Extra cost
Culture	Culture 11 – cities	NGO	Skate tour	Start	Developers	Users	Uncertain sales
Culture	Bonifacius trail	Experts	History	Start	Developers	Users	Uncertain sales
Culture	Claerkamp	Experts	History	Start	Developers	Users	Uncertain sales
Culture	North Art track	Tourist organisation	Wadden Sea	Start	Developers	Users	Uncertain sales
Culture	Wadtours	Experts	Wadden Sea	Start	Developers	Users	No entrepreneur
Culture	Nautic event Fryslan	NGO	Lakes	Start	Users	Users	Uncertain sales
Culture	Model railway museum	Authorities	No special	Active	Authorities	Users	Extra cost
Culture	Belvedere art museum	Authorities	Landscape	Active	Authorities	Users	Uncertain sales
Culture	Heritage accommodation	NGO	Dispersed sites	Active	Companies	Users	Uncertain sales

Appendix

Culture	Steam train Stavoren	Tourist companies	No special	Active	Companies	Users	Extra cost
Culture	Kameleon Terherne	NGO	History	Active	Companies	Users	Uncertain sales
Culture	Peat museum	NGO	History	Active	Users	Users	Uncertain sales
Culture	Oerol festival	NGO	Landscape	Active	Users	Users	Extra cost
Culture	Arts on Ameland	Tourist organisation	Landscape	Active	Users	Users	Extra cost
Culture	Berenloop Terschelling	Tourist organisation	Wadden Sea	Active	Users	Users	Extra cost
Culture	Barents sail days	Tourist organisation	Wadden Sea	Active	Users	Users	Extra cost
Nature	North trail	Authorities	North sea	Start	Authorities	Users	Uncertain sales
Nature	Wadventures	Shipping company	Wadden Sea	Start	Companies	Users	Uncertain sales
Nature	Frisian trails	Experts	Landscape	Start	Developers	Users	No entrepreneur
Nature	Aquazoo	Tourist companies	No special	Active	Companies	Users	Extra cost
Nature	Solar tour Alde Feanen	NGO	Lake	Active	Users	Authorities	Spatial policy
Nature	Night walks	NGO	Landscape	Active	Users	Users	Uncertain sales

Table 3 Sustainable innovations in tourism communication

Communication	Project	Inventor	Quality	Status	Stakeholders	Barrier	Limitation
Physical	Mobile promotion bus	Authority	No special	Start	Authorities	Users	No entrepreneur
Physical	Nodes for biking	Tourist organisation	Culture	Active	Users	Authorities	Extra cost
Virtual	Wadden Sea portal	Authority	Wadden Sea	Start	Authorities	Users	Uncertain sales
Virtual	GPRS for bikes	Tourist organisation	Landscape	Start	Companies	Users	Uncertain sales
Virtual	Goose track	Experts	Bird reserves	Start	Developers	Companies	No entrepreneur
Virtual	Bird watching	Experts	Bird reserves	Start	Developers	Companies	No entrepreneur
Virtual	Panda watching	Experts	No special	Start	Developers	Companies	No entrepreneur
Virtual	Pointer system	Experts	No special	Start	Developers	Users	Uncertain sales
Virtual	Tourism information	Tourist organisation	Dispersed sites	Start	Users	Companies	Uncertain sales
Virtual	Guidance Ameland	Tourist organisation	Dispersed sites	Start	Users	Companies	Product quality
Virtual	Tours Alde Feanen	NGO	Lakes	Start	Users	Companies	Uncertain sales
Virtual	Webcam Alde Feanen	NGO	Landscape	Start	Users	Companies	Uncertain sales
Virtual	SMS guidance	Tourist organisation	Dispersed sites	Start	Users	Users	Product quality
Virtual	Digit art	Tourist organisation	No special	Start	Users	Users	Uncertain sales
Virtual	Digit info in hotels	Tourist organisation	No special	Start	Users	Companies	Extra cost
Virtual	Digit dairy	Experts	No special	Start	Users	Users	Product quality

References

Aernoudt, R. (2008). *Innovation policy and future policy, Belgian and Flanders experiences*. Presentation at the Conference Facilitating Sustainable Innovations, Leeuwarden, June 26–27.

Agarwal, S. (2002). Restructuring seaside tourism: The resort lifecycle. *Annals of Tourism Research, 29*(1), 25–55.

Babcock-Kumish, T. L. (2006). *Trust and antitrust in innovation investment communities, reconsidering moral sentiments*. University of Oxford, mimeo.

Baidal, J. A. I. (2003). Regional development policies: An assessment of their evolution and effects on the Spanish tourist model. *Tourism Management, 24*(6), 655–663.

Baptista, R., & Swann, P. (1998). Do firms in clusters innovate more? *Research Policy, 27*, 525–540.

Bekele, G. W., & Jackson, R. W. (2006). *Theoretical perspectives on industry clusters*. West Virginia University, mimeo.

Cooke, P., & Morgan, K. (2002). *The associational economy, firms, regions, and innovation* (1st ed.). New York: Oxford University Press.

Dwyer, L., Forsyth, P., & Spur, R. (2004). Evaluating tourism's economic effects: new and old approaches. *Tourism Management, 25*, 307–317.

EU (2003). *European Union Regional Policy, Competitiveness, Sustainable development and Cohesion in Europe*; from Lisbon to Gothenburg, European Commission, Brussels.

Faulin, J., Lera, F., Pintor, J. M., & Garcia, J. (2006). The outlook for renewable energy in Navarre. *Energy Policy, 34*, 2201–2016.

Florida, R. (2005). *The flight of the creative class* (1st ed.). New York: HarperCollins Business.

Georg, S., Ropke, I., & Jorgensen, U. (1992). Clean technology-innovation and environmental regulation. *Environmental and Resource Economics, 2*, 533–550.

Gnoth, J., & Anwar, S. A. I. (2000). New Zealand bets on event tourism. *The Cornell Hotel and Restaurant Administration Quarterly, 41*(4), 72–83.

Hardy, A. L., & Beeton, R. J. S. (2001). Sustainable tourism or maintainable tourism: Managing resources for more than average outcomes. *Journal of Sustainable Tourism, 9*(3), 168–192.

Hatakenaka, S., Westness, P., Gjelsvik, M., & Lester, R. K. (2006). *The Regional Dynamics of Innovation, A comparative case study of oil and gas industry development in Stavanger and Aberdeen*. Conference paper, SPRU 40th Anniversary, University of Sussex, Brighton, United Kingdom, September 11–13, 2006.

Hospers, G.-J. (2004). *Regional economic change in Europe: A neo-Schumpeterian vision* (1st ed.). Munster: Volkswirtschaft LIT.

Hunter, C. (1997). Sustainable tourism as an adaptive paradigm. *Annals of Tourism Research, 24*(4), 850–867.

Jorna, R. J., & Faber, N. R. (2006). Sustainability: from environment and technology to people and organizations. In R. Jorna (Ed.), *Sustainable Innovation* (1st ed., pp. 28–41). Sheffield: Greenleaf Publishing.

Kemp, R., Schot, J., & Hoogma, R. (1988). Regime shifts to sustainability through processes of niche formation: The approach of strategic niche management. *Technology Analysis & Strategic Management, 10*(2), 175–195.

Krozer, Y. (2010). Do Innovation Clusters Pay Off ? In M. J. Arentsen, W. van Rossum, & A. E. Steenge (Eds.), *Governance of Innovation* (1st ed., pp. 107–124). Cheltenham: Edward Elgar.

Krozer, Y., & Tijsma, S. (2005). *Sustainable innovations: Using regional qualities for economic development*. Conference of Regional Studies Association, November 24, London, United Kingdom.

Krozer, Y., van de Akker, F., Bijma, T., van der Schaaf, H., & Redzepovic, E. (2005). *Sustainable innovation for tourism mobility*. In 4th international symposium on tourism aspects, mobility and global–local connections, Eastbourne, 23–24 June 2005.

Kruger, L. E. (2005). Community and landscape change in southeast Alaska. *Landscape and Urban Planning, 72*(1–3), 235–249.

Laffitte, P. (2006). *High level advisory group on clusters*. The European Cluster Memorandum, mimeo.

Lanza, A., Temple, P., & Urga, G. (2003). The implications of tourism specialisation in the long run: an econometric analysis for 13 OECD economies. *Tourism Management, 24*, 315–321.

Lawson, C., & Lorenz, E. (1998). Collective learning, tacit knowledge and regional innovative capacity. *Regional Studies, 33*(4), 305–317.

Malecki, E. J. (1991). *Technology and economic development* (1st ed.). New York: Wiley.

Malmberg, A., & Maskell, P. (2002). The elusive concept of localization economies: towards a knowledge-based theory of spatial clustering. *Environment and Planning A, 34*(3), 429–449.

Marelli, E. (2004). Evolution of Employment structure and regional specialisation in EU. *Economic Systems, 28*, 35–59.

Martin, R., & Sunley, P. (2003). Deconstructing clusters: chaotic concept or policy panacea? *Journal of Economic Geography, 3*, 5–35.

Martin, P., Mayer, T., & Mayneris, F. (2008). *Spatial concentration and firm-level productivity in France*. Paris: Centre for Economic Policy Research.

Mitsui, I. (2003). *Industrial cluster policies and regional development in the age of globalisation, Eastern and Western approaches and their differences*. Singapore: ISBC, mimeo.

Moreno, R., Paci, R., & Usai, S. (2005). Geographical and sectoral clusters of innovation in Europe. *Annals of Regional Science, 39*(4), 715–739.

Niosi, J., & Zhegu, M. (2005). Aerospace clusters: global or local spillovers? *Industry and Innovation, 12*(1), 1–25.

Papatheodorou, A. (2004). Exploring the evolution of tourism resorts. *Annals of Tourism Research, 31*(1), 219–237.

Porter, M. (1992). *Concurrentievoordeel* (Competitive Advantage). Amsterdam/Antwerpen: L.J.Veen Uitgeverij.

Porter, M. E. (2000). Location, competition and economic development. *Economic Development Quarterly, 14*(1), 15–34.

Rosenfeld, A. A. (2002). *Creating smart systems: A guide to cluster strategies in less favoured regions, European Union-regional innovation strategies*. Regional Technology Strategies, Carrboro, North Caroline, USA, mimeo.

Scott, A. (1998). *Regions and the World, Economy* (1st ed.). New York: Oxford University Press.

Telfer, D. J., & Wall, G. (1996). Linkage between tourism and food production. *Annals of Tourism Research, 23*(3), 635–653.

Turnock, D. (2002). Ecoregion-based conservation in the Carpathians and the land-use implications. *Land Use Policy, 19*(1), 47–63.

Weaver, D. B. (2005). Comprehensive and minimalist dimensions of ecotourism. *Annals of Tourism Research, 32*(2), 439–455.

Webb, W. (2008, March). Can regional clusters be engineered. *Ingenia* (34), 43–46.

Chapter 6
Arts for Environmental Qualities

Can artists foster valuable but presently unrecognised environmental qualities? Cultural uses of environmental qualities, the natural blends, are discussed. Arts are risky investments because results are unpredictable but an arts service as a nexus of artistic and inventors' skills can be socially beneficial, among others with respect to the environmental qualities. Based on the European Capitals of Culture during 2000–2009 it is shown that the arts services generate net income with public–private funding but the cultural and economic performances per visitor and the net incomes per inhabitant vary across the Capitals of Culture. Business plan for the Capital of Culture in 2018 envisages arts services for the poorly articulated environmental qualities. Innovative uses of environmental qualities are envisaged. Experiences with the arts services using environmental qualities show high leverage of the public funding. The arts services can generate net social benefits based on the unrecognised environmental qualities.

6.1 Arts and Services

Many environmental qualities are lost, their value being recognised after the loss. For instance, the dodo could have been more loveable than the panda if not hunted down and Lebanese cedars could have mitigated desertification of the Levant if not felled for wars and temples. Similar holds for environmental impacts. For instance, the climate change caused by carbon dioxide emissions was issued only thirty years ago; degradation of sea by plastic particles, so-called 'plastic soup', ten years ago, and darkness is still a non-issue. Environmental qualities that could enable new valuable activities may be lost by the time the activities would take off, their high value being recognised after the loss; this also holds for the impacts. Present policies protect areas assumed valuable by the environmental metrics. It implies losses on other areas, but welfare is lost if the metrics are found deficient after a while. The unrecognised environmental qualities have an existence value even if their use is imaginary but how to identify and use such qualities is a challenge. In addition to protection of areas, valuable uses of environmental qualities could be envisioned and recreated for the natural blends as introduced in Chap. 3. The novel natural blends based on the presently unrecognised environmental qualities could be envisaged by individuals who are sensitive to smell, light, sound, space and suchlike environmental qualities and are capable to transform these senses into inventions. Artists are usually recognised for sensitivity to such qualities and imagination, and many artists consider themselves as being inventors. Wassillie Kandinsky, a painter in the early 1900s, expressed it as an artist '…works double hard: (1) he himself must be the inventor of his own art, and (2) he had to master the technique necessary for him. His importance in the history of art can be measured by the extent to which he possesses these two qualities' (Kandinsky (1920) 1989:438). Herewith, it is hypothesised that artists who are interested in environmental qualities can provide services for the novel natural blends. An arts service is considered a nexus of the artistic and inventor skills aiming at a social demand. Whether the arts services can be socially beneficial is discussed using experience of the European Capitals of Culture with particular attention for environmental qualities.

A nexus of arts, environment and innovations is in infancy but there are signs of development. A review of environmental qualities in arts throughout the last few centuries shows the richness of arts expressions (Thornes 2008). The modern environmental arts along with similar labels embrace images, performances, machines and constructions with regard to social activism, philosophy, aesthetics, drama and so on. The works range from technologies of high engineering details to images of global issues (Weintraub 2006, 2007). Scholars follow with a time lag. Environmental studies rarely refer to the arts. A review of studies on the ecosystem services for culture shows growing interests in the cultural expressions of environmental qualities but hardly anything about the arts (Milcu et al. 2013). Art is extensively covered in the cultural economics. Many issues are reviewed, such as the artists' incomes and market, pricing versus public funding (Baumol and Bowen 1965; Heilbrun 2001), patronage systems, freedom of expression, work for private and public interests (Throsby 1994) and productivity and uses of modern technologies (Baumol 2006). Art is also related to sustainable development being considered a system to sustain attitudes and beliefs in society (Throsby 1995), but environmental qualities are untouched. Art is rarely found in the studies of environmental economics though it has low environmental impacts and can be of high value. Compare, for instance, a hundred thousand dollar costing a troupe of five artists that would typically consume 3 tons of materials a year and create valuable natural blends with a team of five construction workers who consume on the annualised basis nine times more materials to build a two-storey concrete family house that irreversibly destroys the site. The development of the arts services aiming at the innovative natural blends is also hindered by deficiencies on the demand side and supply side of the arts.

On the demand side one finds large art consumption but deficient assessments of the arts qualities. The European Union statistics (Eurostat) show that the consumers' expenditures on culture and recreation are 7–12 % of all their expenditures in the low- and high-income European countries, respectively, which is a similar percentage to the health expenditures. A higher percentage is found in Japan and a lower in the United States. The data on culture alone, available only for a few European countries, show 8–15 % of the expenditures on culture and recreation. It means that 1 % to 1.6 % of all consumer expenditures are on arts, which is €102 (USD 133) to €184 (USD 239) per person a year, as much as, for example, on water services. This is sufficiently large to generate the arts services. The difficulty is to assess their quality because markets of arts are myopic. The contingent valuation is advocated in the cultural economics (reviewed in Dupuis 1985; Noonan 2003), but there are unsolvable problems with this approach. One is that the respondents are rarely connoisseurs of arts as they are experts on utensils and their neighbourhood and many people express social conventions about culture rather than personal views (Throsby 2003). Second is the deficient comparison of the willingness to pay for arts from the aesthetical and engagement viewpoints (e.g. Guetzkov 2002; Klaic 2003) versus payments for leisure though it is generally agreed that the arts and fun differ. Third, people are willing to pay after a positive experience of performance but less or not after a poor experience. In effect, artists and art entrepreneurs prefer assessments of the direct costs and benefits because they are tangible to funders though can be biased by political opportunities (Frey 2001).

On the supply side one finds deficient rewarding mechanism. There are about 2.5 million registered artists in the United States and 6.7 million in the European Union (Artists 2014). Most of them are poorly rewarded although investments in arts show positive rates of return albeit lower than many alternatives (Frey and Eicheberger 1995), and arts generate productivity, for example, due to applications in design and media called the 'creative industry' (Scott 2004). The problem is that the inventing artists rarely benefit from their work. Illustrative is that the early realistic paintings of Vincent van Gogh were sold in his time but forgotten by now and his impressionist works were not appreciated by connoisseurs of his time but nowadays attracts about 1.5 million visitors a year to the museum named after him in Amsterdam (Winsemius 2006). Similar holds for the Frans Kafka literature in Prague; Charlie Parker music in Harlem, New York, and so on. Companies also gain but rarely pay. Edgar Varese and Iannis Xenakis, for instance, assisted Philips in research on sound but missed the gains of the firms' inventions. Poor rewards to inventors is not limited to the arts, but in the arts the rewarding is aggravated because qualities are generally assessed by the intermediaries called criticists or connoisseurs, not by peers as it is in science and technology or by users as it is on markets. This intermediary approach causes additional deficiencies because their quality perceptions differ from the peers and public.

6.2 Arts Services Performance

The costs and benefits of the arts services are discussed using experiences of the European Capitals of Culture. These are annual cultural events awarded by the European Commission. The events followed the proposal launched by Melina Mercouri in 1985 aiming to highlight the richness and diversity of the European

Table 6.1 Selected data on Capitals of Culture using Liverpool 2008 framework

Indicators	Year Cities (×1000 inhabitants)	2007 Luxembourg (90)	2007 Sibiu (150)	2008 Liverpool (810)	2008 Stavanger (120)	2009 Linz (190)
Cultural access and participation						
Audience number in million		3.3	2	9.8	2.5	3.5
Satisfied, good, excellent		60 %	90 %	87 %		
Active volunteers, number		241	1200	971	489	
Economy and tourism						
Income from visits in million euros (€)		56	64	740		78
Tourists' number in million		1.21	1	9.7	2.25	2.97
Vibrancy and sustainability						
Income of arts				87.6		8.4
Submitted project		1145	867		1118	2000
Realised projects		584		271		200
Extra-creative businesses				135		93
Extra-creative jobs				879		
Image and perception						
Number of items in media		3380		950	5468	
Media positive %		29 %		56 %	55 %	
Awareness about Capital of Culture		40 %		60 %		60 %
Public positive opinion		90 %	65 %	95 %	78 %	88 %
Government and delivery process in million euros (€)						
Total expenditure		57	17	155	39	67
EU funding		0.8	1.8	1.5	1.4	1.5
Local public funding		13	9	90	15	20
Other public funding		40	4	33	12	40
Private funding		5	2	31	11	7

Garcia et al. (2010) (Liverpool), Richards and Rotariu (2011) (Sibiu), Ecotech (2009) (Luxembourg), McCoshan et al. (2011) (Linz), Rommetvedt (2008) (Stavanger)

cultures within a common history and values. In the subsequent 25 years, two cities are annually selected (once three cities) summing up to 51 Capitals of Culture by 2010. Herewith, the performances of the arts services in five Capitals of Culture of 2007, 2008 and 2009 are compared and the costs and income of all Capitals of Culture between 2000 and 2009 are assessed. The secondary sources are used: the review on the European Capitals of Culture (Palmer 2004), the city-specific evaluations (Garcia et al. 2010; Ecotech 2009; McCoshan et al. 2011) and a few in-depth studies on the specific Capitals of Culture (Greg Richards and Wilson 2004; Herrero et al. 2006; Quinn and O'Halloranc 2006; Rommetvedt 2008; Richards and Rotariu 2011). The assessments differ but better data is unavailable. Table 6.1 shows data for the Luxembourg in 2007 (Luxembourg), Sibiu in 2007 (Romania), Liverpool in 2008 (United Kingdom), Stavanger in 2008 (Norway) and Linz in 2009 (Austria). Vilnius 2009 is excluded because of poor data. The framework of Liverpool 2008 is used.

6.2 Arts Services Performance

The size of the Capitals of Culture varies from 91 000 to 810 00 inhabitants, and the costs and benefits of their arts service vary a lot, as well. The cultural benefits address the audience, opinion and volunteers. The total audience varies from two million persons in Sibiu to ten million in Liverpool but per inhabitant Luxembourg attracted three times more audience than elsewhere. Satisfaction is high except in Luxembourg. Most volunteers are in Sibiu in total and per person. The economic benefits cover income and tourist number. The incomes and numbers of visitors vary from €56 (USD 73) million with 1.2 million visitors to €740 (USD 962) million with nearly ten million visitors. Vibrancy is about the benefits for arts. The largest number of projects is submitted and realised in Stavanger and Linz but per inhabitant in Luxembourg. The arts business, which is known for Liverpool and Linz, grew more in Liverpool than in Linz in total but per person more in Linz. The media coverage is larger in Stavanger and Luxembourg than in Liverpool but it is more critical. The inhabitants' awareness in Luxembourg is low but satisfaction high despite much and critical media. Sibiu has the most critical public. The governance shows the costs. The costs are low in Sibiu but large in Liverpool and Luxembourg in total and per inhabitant. The share of private funding in total varies from 2 % in Sibiu to 31 % in Liverpool. Given the budget, the arts services in Luxembourg have generated more projects but not more visitors and Liverpool with twice larger budget has not realised more projects and businesses. Budget to cover costs is important but not decisive for the benefits. The economic benefit measured by income from arts, known for Liverpool and Linz, is only 11–12 % of all income. This share is similar to the European consumer expenditure, which suggests that the European Capitals of Culture did not generate much benefit for the arts but rather for leisure.

The costs and income are assessed for 18 European Capitals of Culture during 2000–2009. These are Reykjavik, Avignon, Bologna, Helsinki, Rotterdam, Porto, Salamanca, Bruges, Graz, Lille, Genoa, Cork, Luxembourg, Sibiu, Stavanger, Liverpool, Linz and Vilnius. Table 6.2 shows data on the totals, per inhabitant and per visitor. Horizontally one finds: inhabitants, tourists visits, income gross and income net after subtraction of the running costs (not capital costs), infrastructure investments and management staff. Vertically are shown: number of cities, averages, variety indicated by standard deviation and spread shown by the lowest and highest data. The average size of cities was 0.4 million inhabitants and average 3 million visitors with moderate variety but large spread because of a few large cities. Similar holds true for the average of 53 staff involved during season peak. Investments in infrastructure are huge in Salamanca (Spain) and Liverpool that created new art centres. Hence, the average expenditure of €47 (USD 61) million does not vary much but the spread is huge because of these two Capitals of Culture. Most cities get much national funding but little private funding. It is mainly used for the cultural programming. The average income was gross €157 (USD 204) million and net €97 (USD 126) million after subtracting the costs; excluding Salamanca and Liverpool with much infrastructure, it was €71 (USD 92) million and €25 (USD 32) million, respectively. A visit has generated on average €15 (USD 19) net income. There are average 12 visitors to the Capital of Culture per inhabitant but the spread is 6 times

Table 6.2 Performance indicators of the European Capitals of Culture; net income is after subtracting all running costs (payments to artists, organisation and suchlike)

Persons, or euro in the year of the Capital of Culture	Number of cities	Average per city	Standard deviation	Lowest value	Highest value
Inhabitants in million	18	0.4	0.3	0.09	1.1
Tourists in million	18	3.0	2.5	1	9.7
Staff number[a]	13	53	29	10	100
Infra in million euros (€)[b]	12	462	1122	10	4000
Total expenditures in million euros (€)	18	47	33	8	155
From private	10	18 %	14 %	0 %	36 %
From local	10	36 %	15 %	18 %	58 %
From national	10	46 %	24 %	2 %	80 %
For programme	15	68 %	8 %	51 %	80 %
For marketing	7	16 %	5 %	14 %	21 %
For overhead	7	14 %	4 %	11 %	19 %
Income, gross and net (net income = total income − total expenditures)					
Gross income in million euros (€)	9	157	222	41	740
Net income in million euros (€)	9	95	187	−1	585
Gross income in million euros (€)[c]	7	71	23	41	116
Net income in million euros (€)[c]	7	25	22	−1	57
Per inhabitant					
Tourist visit number	18	12	8.0	3	19
Expenditures (€)	18	194	147	52	639
Gross income (€)	9	523	302	112	1059
Gross income (€)[c]	7	390	167	112	627
Net income (€)	9	264	308	−12	824
Net income (€)[c]	7	118	120	−12	309
Net income per visitor (€)[c]	7	15	16	−0.3	46

[a]Number during season peak
[b]Investments in infrastructure
[c]Excluding Liverpool and Salamanca

across the Capitals. The expenditures per inhabitant are average €523 (USD 680) with nearly 5 times spread. These are excluding Salamanca and Liverpool about 390 (USD 507) per inhabitant. It is more than twice higher than the European Union average expenditures on culture per person. All Capitals of Culture except Luxembourg have generated net income. These arts services have generated additional income per inhabitant.

The scale of cities has disadvantages. Based on nine cases with sufficient data, it is found that the unit cost per visitor increases when the visitors' numbers increase. The larger cities have generated less visitors per inhabitant and their expenditures per inhabitant are higher (negative correlations of the inhabitants and visitor per inhabitant with the expenditure per inhabitant, $R^2 = -0.46$ and $R^2 = -0.52$). The medium- and small-size European Capitals of Culture have generated more visitors and operated more cost effectively than the large ones. One would expect that the Capitals of

Culture learn from the previous ones to perform better or to be cost effective. It means that the number of visitors per inhabitant would grow and expenditure per visitor would decrease throughout 2000–2009. There is only little progress in attracting visitors and gross income per visitor ($R^2=0.29$ and $R^2=0.25$) but the performances became less cost effective because more is spent per visitor. The net income per visitor has decreased ($R^2=-0.23$) if the Liverpool and Salamanca are excluded and even worse including these two cities. The visitors have spent more throughout these years but the cities' expenditures increased even faster. The willingness to pay seems to increase but the arts services must bear even more costs to attract visitors. The 'cultural hangover' when budgets for culture are gone after the event is a risk. The arts services of Capitals of Culture have generated more income with the public funding but rewards for the artists are low because nearly 90 % of all income is spent on leisure and continuity in arts is at risk when budget is spent after the events.

6.3 Arts Service of Natural Blends

The arts services that use various environmental qualities are illustrated based on the Leeuwarden European Capital of Culture 2018 in the Netherlands. Several arts services are briefly introduced followed by an economic assessment of one of the services. The European Commission proposed Malta and the Netherlands to be the European Capital of Culture in 2018. In the Netherlands, Leeuwarden won the competition with Utrecht, Maastricht, The Hague and Eindhoven. The Leeuwarden bid book is focused on the regional environmental qualities and social capabilities for cultural development. These activities cover three themes, each one with three main projects and many supporting events. Herewith, only the main projects are mentioned. More can be found on http://www.2018.nl/.

The theme nature and culture addresses the biodiversity loss, which threatens the cultural diversity. Links between culture and nature are shown. The 'Sense of place' refers to the Oerol festival. It is an outdoor festival of location theatres on an island on the Wadden Sea (UNESCO World Nature Heritage). The performances take place outdoors using the natural and man-made environmental qualities of the island: the dykes, meadows, forests, beaches, dunes, sea and so on. A site for the international performances on the island is proposed with continuation after 2018. The 'Embassy for water' is an art project with artists, ecologists and scientists who express various dimensions of water: still, running, sea, air and so on. This project relates to water technology development in the existing Water Science Park. The 'Spring Fever' is a travelling theatre that reflects on threats for life of the godwit and bees on their routes. The arts related to the godwit life will reconnect people from different countries on the godwit route and artistic events are envisaged to celebrate pollination. The city and countryside theme is focused on creativity and ecology. Its aim is to close the gap between urban and rural communities through the novel use of environmental qualities. 'Feel the Night' envisages the Dark Sky Park where visitors can sail on electric boats to experience darkness and silence in a national park. Art performances with sky observations in the night aim to strengthen the sense of darkness and silence. An educational observation centre is considered for the follow-up. 'Farm of the World' is an outdoor atelier for artists to live and work on a farm. Use of the local resources for arts is envisaged. This atelier will be linked to portable farming, urban vertical roof gardens and novel farming methods. The 'Eleven Fountains' project shows various possibilities for running water in open space. Fountains situated in landscape underline the specific local conditions. Glass and laser technology supports art exhibits about fluids. The Community and Diversity is about how people of different cultures live together. The 'Language Lab' is the multilingual museum that uses crowdsourcing for the linguistic surveys and shows the diversity of tongues with modern technologies. The 'Strangers on Stage' is a theatre festival about the intercultural diversity, conflicts and dialogue. It involves performances and masterclasses in cooperation with theatres abroad. The 'Lost in the Greenhouse' is a stage and a theatre for exploring experiences of the immigrant workers. The multicultural crossovers with theatres abroad in greenhouses are envisioned. These arts expressions can generate innovative natural blends related to the bird migration, pollination, sense of silence and darkness though follow-up products and services are still unclear.

The Oerol festival illustrates that such uses of environmental qualities can generate income for artists. The Oerol costs and income are assessed in 2005 based on the regional monitoring, festival evaluations and interviews with entrepreneurs, policy-makers, artists and social organisations on the island where this festival takes place (Krozer 2003). The festival duration is 10 days but most visitors of the festival stay for 4 days on the island. The regional tourism monitor estimated 23,000 visitors to the Oerol based on inquiries but the festival organisation estimated about 35 000 visitors using the tickets sales. This difference cannot be explained by the tickets sold to the inhabitants on this island because only a few thousand persons live there; the inquiries could be less precise than the ticket sales. Visitors to the festival cover about 15 % of all visitors on the island throughout the year. These visitors highly value the island (8.4 out of 10) and the festival (8.5 out of 10). The number of repetitive visits to the festival is high. Stakeholders on the island are less positive. The inhabitants value it lower (6–7 out of 10) and one-third complains about nuisances, though one-third of inhabitants support the festival with money for the festival guarantee fund. Nearly half is inspired by the festival and is positive about the visitors' behaviour, and about two-third visits a performance. The entrepreneurs are even less positive. The chairman of the local entrepreneurs' association argues that the festival generates low income compared to the number of visitors but most entrepreneurs on the island contribute to the guarantee fund. The total costs of the festival were about €1.96 (USD 2.5) millions. About half of the income is covered by entry tickets, nearly one-third by expenditures on leisure and one-fifth by public funding. The direct income from those visitors who stayed on the island with the purpose to visit Oerol is estimated to be €6.7 (USD 8.7) millions. This includes trips, hospitality, leisure and other local services. About 29 % of it is spent on the festival, which is twice higher percentage than in the Capitals of Culture of Liverpool and Linz (more data is not found). The public funding invoked 11 times larger income due to combination of art, fun and environmental qualities of the island.

6.4 Conclusions

Innovation strategy is advocated for valuable environmental qualities that are not recognised, yet. The strategy is that the arts generate services that envision and develop the valuable uses of environmental qualities, the natural blends. The consumers' expenditures on arts are sufficiently large for such arts services but the demanded qualities are difficult to assess and fair rewards for artists difficult to reach. The arts services performed during the European Capitals of Culture in the period 2000–2009 show cultural, social and economic benefits given costs albeit performances vary. The average income per inhabitant exceeds the European average more than twice but only about 12 % of this income is for the arts and the remaining part for leisure. The hospitality industry is the main beneficiary. Infrastructure increases costs of these arts services without sufficient income increase. The large-scale activities have negative effects on the net income. There is also no much learning about making such arts services cost effective and discontinuity after events poses a threat to the artists' income. The proposal for the European Capital of Culture Leeuwarden 2018 shows that artists are interested in using environmental qualities for cultural expressions. A nexus of arts, environment and technologies is envisaged with uses of darkness, silence, openness, life of animals, local nature and other environmental qualities for cultural activities. The arts services that combine the artists' and inventors' skills can generate beneficial activities based on the poorly recognised environmental qualities. Linking artists and inventors for envisioning environmental qualities is socially beneficial with potential high rewards for artists.

References

Artists. (2014). For the European Union is http://epthinktank.eu/2013/04/17/european-cultural-creative-sectors-as-sources-for-economic-growth-jobs/, for the United States is http://www.princeton.edu/culturalpolicy/quickfacts/artists/artistemploy.html. Visited 21 Dec 2014.

Baumol, W. J. (2006). Arts in the "New Economy". In V. A. Ginsburgh & D. Throsby (Eds.), *Handbook of the economics of art and culture* (Vol. I, 1st ed., pp. 340–357). Amsterdam: Elservier.

Baumol, W. J., & Bowen, W. G. (1965). On the performing arts: The anatomy of their economic problems. *American Economic Review, 55*(1–2), 495–502.

Dupuis, X. (1985). *Application and limitation of cost-benefit analysis as applied to cultural development*. Paris: UNESCO, mimeo.

Ecotech. (2009). *Ex-post evaluation of 2007 & 2008*. European Capitals of Culture, Brussels, mimeo.

Frey, B. S. (2001). Public support. In J. Heilbrun & C. M. Gray (Eds.), *The economics of arts and culture* (1st ed.). Cambridge, UK: Cambridge University Press.

Frey, B. S., & Eicheberger, R. (1995). On the rate of return in the art market: Survey and evaluation. *European Economic Review, 39*, 582–537.

Garcia, B., Melville, R., & Cox, T. (2010). *Creating an impact: Liverpool's experience as European capital of culture*. University of Liverpool, Liverpool, mimeo.

Greg, R. G., & Wilson, J. (2004). The impact of cultural events on city image: Rotterdam, cultural capital of Europe 2001. *Urban Studies, 41*(10), 1931–1951.

Guetzkov, J. (2002). *How the arts impact communities*, mimeo. Princeton University. Princeton. https://www.princeton.edu/~artspol/workpap/WP20%20-%20Guetzkow.pdf. Visited 15-9-2014.

Heilbrun, J. (2001). Baumol's cost desease. In J. Heilbrun & C. M. Gray (Eds.), *The economics of arts and culture* (1st ed.). Cambridge, UK: Cambridge University Press.

Herrero, L. C., Sanz, J. Á., Devesa, M., Bedate, A., & del Barrio, M. J. (2006). The economic impact of cultural events, a case study of Salamanca 2002, Europe capital of culture. *European Urban and Regional Studies, 13*(1), 41–57.

Kandinsky, W. ((1920) 1989). *Complete writings on arts* (1st ed.). New York: De Capo Press.

Klaic, D. (2003). *Performing change, debating the issues*. How could performing arts strengthen the public debate about food, animals, landscape and the farmers, Amsterdam, mimeo.

Krozer, J. (2003). *Regionale economie van cultuur in Fryslan*. Advies, 30 November, 2003. Amsterdam: Mimeo (Available on request).

McCoshan, A., Rampton, J., Mozuraityte, N., & McAteer, N. (2011). *Ex-post evaluation of 2009. European Capitals of Culture*. Birmingham: Ecorys, mimeo.

Milcu, A. I., Hanspach, J., Abson, D., & Fischer, J. (2013). Cultural ecosystem services: A literature review and prospects future research. *Ecology and Society 18*(3), 44. http://www.ecologyandsociety.org/vol18/iss3/art44/

Noonan, D. (2003). Contingent valuation and cultural resources: A meta-analytic review of the literature. *Journal of Cultural Economics, 27*, 159–176.

Palmer, R. (2004). *European cities and capitals of culture*. RAE Associates, Part I and II. Brussels: Mimeo.

Quinn, B., & O'Halloranc, I. (2006, November). Cork, 2005, An analysis of emerging legacies, Mimeo.

Richards, G., & Rotariu, I. (2011). *Ten years of cultural development in Sibiu: The European cultural capital and Beyond*. Munich Personal RePEc Archive. Munich: Mimeo.

Rommetvedt, H. (2008). *Beliefs in culture as an instrument for regional development: The case of Stavanger*. European Capital of Culture 2008, conference, ERSA. Liverpool: Mimeo.

Scott, A. J. (2004). Cultural products industries and urban economic development. *Urban Affairs Review, 39*(4), 461–490.

Thornes, J. E. (2008). Rough guide to environmental art. *Annual Review Environmental Resources, 33*, 391–411.

Throsby, D. (1994). The production and consumption of the arts: A view of cultural economics. *Journal of Economic Literature, 32*(1), 1–29.

Throsby, D. (1995). Culture economics and sustainability. *Journal of Cultural Economics, 19*, 199–206.

Throsby, D. (2003). Determining the value of cultural goods: How much (or how little) does contingent valuation tell us. *Journal of Cultural Economics, 27*, 275–285.

Weintraub, L. (2006). *Eco centric topics, pioneering themes for eco art* (1st ed.). New York: Art Now Publications.

Weintraub, L. (2007). *Environ mentalities, twenty two approaches to eco-art* (1st ed.). New York: Art Now Publications.

Winsemius, P. (2006). Lecture at the opening of the Cartesius Institute. Leeuwarden, 31-8-2006.

Chapter 7
Technology Suppliers for Sanitation

Can lock-in into a costly technology be resolved? The dominant and alternative networks of interests in sanitation technologies for households are presented. The policy-steering role of the sanitation entailed technological lock-in into a lengthy transport of excrements in sewage and multistage wastewater treatment, which is effective but costly, even unaffordable to a large part of the global population. The alternatives to this dominant elongation system are cost-effective, but each one has drawbacks. One alternative network pursues separation at sources, which is cheaper but needs behavioural changes at home. The distributed sanitation treatment is close to homes, which reduces the transport and treatment costs, but it needs space. Finally, the value-adding services pursue uses of excrements, which can be net beneficial but less effective than the dominant network. Policies that tune to the communities' needs and capabilities can resolve the lock-in with more socially beneficial innovations.

7.1 Maxi–Min Regulation

Each technology supplier involves a network of interests that delivers technologies and services. Adoption of a technology implies embracing a network of interests whose supplies are tuned to each other by their quality, performance and price. A side effect is that firms in the network whose services are not chosen are out of this business unless they can tune to adopted technology network for their future supplies or they create a rival network. When the dominant technology network is vested other suppliers must tune to it. A singular technology adoption could be replaced by the superior alternative whenever the cost of the installed technology is sunk, or the present value of its remaining life time is higher than the present value of the alternative. 'Bygone is bygone' is the conventional economic teaching about replacement. Replacement of a network, however, poses the risk of quality deficiency because some supplies in the network may not match (other firms gladly sell connectors, batteries and suchlike due to the network deficiencies). The dominant technology networks, therefore, are usually maintained even if a new technology is superior because substitution is risky. The dominant technology network evolves into a lock-in of vested interests. Its costs can grow faster than the cost-saving technological progress but the lock-in impedes innovations.

The lock-in evolves spontaneously on markets but policies can reinforce this process or induce alternatives. Presently, the lock-in is usually induced by policies, particularly with respect to environmental qualities. When the social pressures for good environment have reached the political decisions, policymakers generally consult experts and firms about technology suppliers that reduce environmental impacts. The technologies are usually selected based on the principle 'As Low As Reasonably Achievable' (ALARA) and suchlike criteria. These technologies must be proven and broadly accepted by experts and industries called 'Best Available Technologies' (BAT) or similar. This policymaking is statutory in many countries, for instance, in the European Union. The aim is maximise effect (ALARA) at minimal risks (BAT). This is maxi–min strategy in policy decisions. After this policy preparation the selected technologies are enforced using permits, charges, public investments, moral suasion and other instruments but the policy evaluations a few decades after the policy preparation often show deficient performance. The decision makers can be blamed for these deficiencies as being too stringent or risk avoiding at the time of decisions though they can be right from the institutional perspective of risk avoidance (March and Olsen 1984). The problem is that such maxi–min strategy vests interests that evolve in a technology network entailing the lock-in.

The conventional policymaking is deficient but alternatives are not better. The technology forcing, meaning stricter policy demands than the demands achievable using the BAT technologies, has a risk of failures. It fails when the demanded performance is not attained. Misjudgements cannot be avoided because the successful technologies in long term are unpredictable, and the scaling of cost and effects of the past technologies for the different future uses is unreliable (Krozer 2006). Market negotiations about environmental performance are usually even less cost-effective and delay decisions because they are laborious and end up on the policymakers' desks entailing the conventional decision making with high transactions costs. For example, in the Netherlands such negotiations about pollution reduction sealed in more than 100 covenants in the 1990s caused costlier and less effective policy and nearly doubled the transaction costs. These negotiations enable laggards to delay actions at the social costs (Krozer et al. 2013a).

Herewith, the diversity policy is advocated to resolve the lock-in. In the diversity policy alternative criteria for decision making about technologies can be used in relation to the communities' needs and capabilities, for example, accept higher risks at lower cost. This possibility is shown for the household sanitation (Krozer et al. 2013b). Different criteria in the communities' sanitation policy can induce other technological networks. Four sanitation technology networks are illustrated. The dominant sanitation is elongation through adding technologies to each other, which is based on the maxi–min strategy. It is effective but costly. An alternative is the separation at sources of wastewater, which is cost saving but needs self-work.

Another option is the distributed sanitation with constructed wetland. This reduces capital costs of infrastructure but needs space. Finally, the value-adding services can be net beneficial but less effective. The aims of all these methods are similar: prevent threats for human health (e.g. diarrhoea, cholera, typhoid, hepatitis), satisfy social needs (e.g. leisure, tourism, potable water, industry water) and attain healthy ecosystems (e.g. rivers, wetlands, mangroves, sea) with regard to pollution from toilets, kitchens and showers, companies' process water as well as run-off from land, roads and roofs. Nevertheless, the alternatives differ if measured by the physical and monetary indicators.

7.2 Conventional Sanitation

The presently dominant sanitation is a nineteenth-century innovation. When the sewage constructions emerged shortly after the French revolution (1789) in Paris, they substituted vaults and cesspools under houses filled with chemically treated wastewater that soaked into ground or were brought away. A privy on backyard was rare; a septic tank is invented in the late 1800s. Wherever sewage is introduced, the water-borne epidemics decreased. Sewage has been a major innovation for social health in centuries. As usual also this innovation met fierce opposition of the vested interests. Whenever it was introduced in France, the United Kingdom, the United States and other industrialising countries, the house owners who were unwilling to pay a fee and the farmers who were claiming dry dung for soil enrichment amalgamated their interests into a force that obstructed sewage constructions. The obstructions went on throughout the 1800s despite regular outbreak of cholera, typhoid and so on. Only by 1887 the Prefect of Paris, Eugéne Poubelle, was able to enforce sewage with fees on the house owners; his name is engraved in the French language for waste. Sewage expanded mainly during the economic upswings of the 1910s and the 1960s in the United States, Europe and Japan when capital was abundant to cover large infrastructural investments because sewage is capital intensive (Beder 1990; Rockefeller 1997; Burian et al. 2000, Paris project 2013). Still in the 1990s only about 40 % of the global population was connected to sewage and even by 2010 only 63 % are connected: 30 % in the sub-Saharan Africa, about 60 % in Asia, up to more than 80 % in Latin America and 90 % in the OECD countries. It implies only 2 % average annual rate of technology dissemination, which is several times lower than the rate on the markets of electricity, roads and telecommunication. The countries' policies also differ. For example, Sri Lanka has more sanitation per person than twice richer per person in Mexico. Mostly urban areas are served, but slums are rarely sanitised; 55 % of the rural population lacks sanitation. Mainly the rich and middle class are served, but they rarely pay all costs. Hence, sewage deteriorates. The untreated wastewater discharges into ambient water are prohibited, but they occur when corrupted pipes leak or municipalities have no treatment at the end of the pipe (UNDP 2006; WHO/UNICEF 2012).

In the course of the last century, the sanitation methods that were introduced in France have disseminated across countries and elongated within the countries to cover more households and businesses including the dispersed ones in rural areas. These generated various technology supplies. More connectivity exponentially enlarged the pipe length and the transport mileage involving stronger materials for pipes, more power for pumping and better monitoring of operations. Double sewage is in development with the aim to separate drainage from pollution flows and to cope with the polluted run-off during heavy rainfalls. Wastewater treatment that emerged early 1900s became more effective through adding technologies to each other. It is elongated into a multistage treatment. A minimal treatment is the mechanical separation and sedimentation of solids in septic tanks or basins (1st stage). It is often followed by the more advanced treatment using the biological degradation under oxygen-free conditions (anaerobic). Then, the oxygen-enriched (aerobic) treatment with activated sludge technologies is increasingly applied (2nd stage). Some wastewater plants also treat the specific pollutants using various biological and chemical technologies for nitrates, phosphate and heavy metals (3rd stage). An emerging stage is the so-called 'polishing' which aims to reduce the microbial, viral and hormonal pollution. Sediment that is accumulated in sewage (silt) and in wastewater treatment (sludge) consists of dust, sand, minerals and organic matters. It is usually collected and stored on farmland, or it is landfilled sometimes after digestion with biogas winning. This elongation has reduced risks for public health and water ecosystems (Cooper 2001). The elongation became the dominant technology network. It is effective but costly.

Studies for the WHO suggest that 2.1 billion people without sanitation can be served through USD 200–260 (€154–200) investments per person entailing USD 9–15 (€7–12) annual costs. These costs would be two to five times lower than the social benefits of better health, productivity and so on (Hutton and Haller 2004; Haller et al. 2007; Hutton and Batram 2013); the benefits for environmental qualities are not covered. Based on these studies the World Health Organization argues that the present USD 115 (€88) billion global annual expenditures on the sanitation should increase by USD 25 (€19) billion a year to meet the Millennium Development Goals that aim to close the gap between the low-income and high-income countries among others in sanitation in 2020 as agreed at the Global Summit on Sustainable Development in 2000 (WHO 2012). The costs, however, are underestimated if the elongation network is pursued because investments in sewage increase exponentially with connections, which entails higher annual capital costs and involves higher operational costs because of more maintenance and energy use. The costly multistage wastewater treatment is also necessary and adds to the costs. The annual costs could be up to ten times higher than the projections above if the aim is to serve effectively more than 90 % of all people. This is needed but high connectivity and effective treatment are costly.

Experiences in the high-income countries show that an advanced elongation network for a 100,000-inhabitant city needs typically USD 130 (€100) million investment costs, which involves about USD 15 (€11) million expenditures a year paid by the inhabitants and regional and national funds. Since it is politically uneasy to ask people a few hundred dollars annual payment without options for income generation, the modern sanitation disseminates slowly (Krozer et al. 2010). Take the Netherlands as example of the high-income countries. During the period 1985–2004 its population has increased by 12 % to nearly 16 million people, but the annual costs of sewage have increased by 288 % to about USD 1300 (€1019) million in 2004 mainly because dispersed houses in the rural areas are connected. The multistaged wastewater treatment caused more than a tenfold unit cost increase from nearly USD 0.1 (€0.07) per m^3 wastewater in the 1st stage to more than USD 1.0 (€0.8) per m^3 for the 2nd and 3rd stages. The costs of silt and sludge disposal also increased from a few dollars per ton disposal on farmland to a few hundred dollars per ton for the obligatory landfilling. The digestion for biogas before the landfilling became economic mainly because it reduces the volume of disposed sludge. The sanitation costs in the Netherlands still increase.

The investment cannot be reversed during decades. In the annualised life cycle costs of water supply and sanitation, about 70 % is depreciation on past investments, out of this 50–60 % for pipes and the remaining for treatment, of which 40–50 % for construction and similar amount equipment, whereas the remaining 10–20 % is know-how. About two-thirds of the annualised costs are fixed based on 25 years of depreciation. The remaining costs are for the staff (about two-thirds of these) and for chemicals, energy, space and others (about one-third). The high capital costs imply that changes are costly and risky because scrapping for an alternative network involves numerous suppliers whose different interests and qualities must be tuned to each other. Many technological improvements within the elongation networks are introduced aiming to increase the effects and reduce costs in one or in a few steps, but these hardly improve the overall effects but increase the costs. The elongation network has effectively reduced the negative health and environmental impacts of excrements but at high and increasing costs.

7.3 Sanitation Alternatives

The alternatives cover only a few percent of the global sanitation market.

7.3.1 Separation Network

Instead of bulking all excrements into sewers, the separation networks are focused on catching residues in wastewater for reuse at source (reviewed in Krozer et al. 2010). This can generate high environmental quality when water is saved and reused and matter in wastewater is used for biofuel and minerals. It is also cost-effective if the price of water use is high and the technologies for separation and reuse of matter are low cost. The difficulty is that the separation requires major changes in housing and in people behaviour. The separation technology is usage specific, which means that the household and professional activities must pay attention to the specific uses. Expertise presently alien to sanitation is needed, for instance, architects for adaptations of houses, designers for water-saving tools, social work for behavioural changes. A reward is that 20–25 % of all water use in households in the high-income countries can be cost-effectively saved. Less water use means lower discharge of wastewater and less costly sanitation because less power is needed for wastewater transport and less wastewater is treated. The reuse of matter saves some costs.

The main water usage and savings are due to toilet flushing. For instance, a typical toilet flush in the Netherlands or Germany takes about 5 L of water compared to standards 9 L in the United Kingdom and 12 L in the United States. The dual-flush toilets give the user the option of a 3 or a 6 L flush. Experiments with the dual-flush show about 27 % saving. There are many flush-saving devices, such as delayed action floats, syphons containing air and drop-valve mechanisms connected to infrared sensors. The vacuum toilets in trains, ships and airplanes and in some households in experimental settings use air pressure and about 1 L of water per flush. The composting toilets use almost no water because waste is decomposed in a container below the toilet; types vary from the compact one-unit toilets to advanced multi-tank options with devices to enhance decomposition by solar units, computerised monitoring and combinations with kitchen waste. Much water is used for bathing and washing. Efficient showerheads and taps also reduce water use. Experiments show nearly 17 % water saving when such devices are used. Water-saving dishwashers and washing machines also reduce discharges along with less energy consumption. Rainfall is often used in Asian countries for the non-potable usages, such as toilet flushing, air conditioners and gardening, but rarely in Europe. Experiments with water reuse from showers, washing machines and dishwashers for toilet flushing and gardening were successful in arid regions in Australia and the United States. The low scale of operations indicates high barriers for the separation networks. There are technical barriers, for instance, space in houses. The behavioural and legal issues reflect public concerns about health risks. The public attention is also low because water is perceived abundant and sanitation low priority.

7.3.2 Distributed Network

Biodegradation of excrements can be done in nearby housing instead of transporting them far away. Septic tanks are used for the collection and sedimentation of suspended solids followed by biodegradation with use of the wetland plants (helophytes), such as reed, sweet flag, bulrush and others. Such distributed sanitation is cost-effective compared to the elongation and safe if the treatment is located under the soil surface. The distributed sanitation based on these constructed wetlands reduce the annual costs of constructions, maintenance and energy use for transport. The disadvantage is the large-area use. The space use that is necessary for adequate wastewater treatment can be derived from the BOD removal kinetics. It is (after Arceivala and Asolekar 2008):

7.3 Sanitation Alternatives

$$C_t = C_0 e^{-Kt} \tag{7.1}$$

For

$$A = f(t) \tag{7.2}$$

or

$$A = \frac{Q(\ln C_0 - \ln C_t)}{K_{BOD_5}} \tag{7.3}$$

where bed area is in m², A is the logarithmic function of the average flow in m³/day, Q is the inlet 5 day-BOD in mg/l, C_0 is the outlet BOD_5 in mg/l, and K_{BOD_5} is the reaction constant per day.

The K factor, estimated empirically, varies from 0.067 in cold and wet in the United Kingdom and 0.083 in Denmark and up to 0.17 in warm Bangalore, India. In general, the K factor is assumed a logarithmic function of temperature. Up to twice higher values can be reached in warm humid countries. Given high environmental standards, the horizontal flow constructed wetlands need 1.0–2.0 m² per person equivalent wastewater and the vertical ones 0.8–1.5 m² in the moderate-climate countries (UN-HABITAT 2008: 19). Low stringent standards and warm climates reduce the space. Better dosing of wastewater on the wetland area also improves their performance and reduces the space use. Cost-effective innovations can be aeration of soil to enhance the biodegradation and substitution of septic tanks for dispersion but these are not proven technologies. Even very effective constructed wetlands consume a few times more space than the advanced multistage wastewater treatment plants that need about 0.15 m² per person, and the bio-membrane installations are even denser but costlier. The advantage for the low-income countries, in small towns and in rural areas is that the construction can largely be executed and operated by the local firms, farmers or volunteers if space is available. The constructions and operations can be found in manuals that show step by step how to construct and operate such technologies (e.g. Hydrik 1998; Tousignant 1999; EPA 1999; UN-HABITAT 2008). A few main steps in the construction and operations of the vertical flow constructed wetland are shown in the Picture Serial 7.1, which is based on the BrinkVos Water experiences with a few thousand installations at household, farms, industries and landfill.

The serial shows: (a) excavation of soil; (b) soil protecting impermeable foils; (c) layers of shells, gravel and clean sand; (d) pipes for drip irrigation under soil surface and sowing reed and (e) connection of toilets to septic tanks (not shown on the pictures). During operations wastewater is settled and digested in septic tanks with biogas production as a side effect, then pumped on top of the filter bed and spread through pipes under the soil surface to drip into the filter. This activates biodegradation

Certification data for the BrinkVos vertical flow helophytes filter in mg/l; it is the highest possible: IIIB, certification		
parameter	Certification criteria	Realized average
COD	100	17
BOD	20	3
N - Total	30	18.2
NH_4-N	2	0.4
P – Total	3	0.09
TSS (suspended solids)	30	2.2

Picture Serial 7.1 Construction of the BrinkVos water vertical flow helophyte filter on 400 m^2 for a school with 1000 children and 150 households in Culemborg; from upper left, clockwise: excavation 1 m deep; foils, shells, lava; clean sand filling; drip pipes and reed. Beneath the average results certified by KIWA are shown. I am grateful to Tinus Vos

by microorganisms at the plant roots with absorbance of minerals to sand, seashells and gravel. Grown reeds are mowed; good results in warm climates are with *Cyperus* and *Miscanthidium* (Kyambadde et al. 2004). Water after treatment is controlled and can be reused for toilets and gardening. The externally certified performance shows that the effluent quality after this treatment exceeds the criteria for disposing effluents to ambient water in the Netherlands. These criteria are among the most stringent in the world. The constructed wetlands cannot reach the footprint of the advanced wastewater treatment plants. They satisfy the maxi–min policymaking and reduce infrastructure, maintenance and energy, but the costs of land use in urban areas are high. The large-area coverage pushes this alternative to the margins of sanitation markets.

7.3 Sanitation Alternatives

Parameter	In Septic tank mg/l	Out Discharge mg/l (Dutch norms)	Removal %	Loading kg/ha/d
BOD	145	17.6 (20)	87.9	32.1
Total Phosphorus	8.05	1.9 (3)	76.4	1.7
Total Nitrogen	47.6	10.0 (30)	79	10.3
Total Susp. Solids	69.9	*38.9 (30)*	*44.4*	
Coliform bacteria	49 x 10^6	2.2 x 10^3	99.8	

Picture Serial 7.2 Wastewater Gardens (publications of the Wastewater Gardens Network)

7.3.3 Value-Adding Services

The value-adding wastewater services also use plants for wastewater treatment but creation of value-adding services is the key feature. A trailblazing work is done by the Wastewater Gardens, a network established after experiments with artificial closed ecosystems (Biosphere 2) in the United States; the experiment is stopped because of various difficulties, but it generated innovative approaches, among them concepts of valuable service using wastewater treatment. Picture Serial 7.2 shows gardens in Mexico at a hotel (Nelson et al. 2013 web and Nelson et al. 2006).

The treatment is mainly in septic tanks followed by the sediment discharge and the effluent reuse for the garden. The wastewater garden is only one among many possible services. A web-based search has delivered 55 examples of the value-adding and cost-saving services. The wastewater from households, tourist centres, institutions and run-off is treated. Most examples are in rural areas based on wetlands. The effluent is mainly reused for gardening. Other services are training and education, leisure, biodiversity fields and landscaping, fish ponds, parks, social inclusion projects and so on. Benefits are water saving, groundwater protection, biodiversity in parks and gardens, nature experience on footpaths and walk trails, products for arts with woodworks, fishing in ponds, rain harvesting and so on. The cost saving is mainly due to the water and organic matter reuses. The income-generating services are, for instance, biofuel products, fees for entry to the biodiversity gardens, recreation, fishing, boating, wood carving and water storage in dry areas. However, not many projects pay attention to the income generation and cost saving. Only a few projects made cost–benefit assessments; the benefits of some projects would

Table 7.1 Ideas for linking constructed wetlands with community

Wastewater filter attribute	Activities	Specifications
Space use	Playground	School projects, insight into filtering processes
	Parks	Biodiversity, zoo
	Gardening	Flowers, feed crops
	Ceremonial ground	Spreading ashes, burial
	Adding solar power	Recharging mobiles, sanitation facilities
Rooftops	Urban gardening	
	Climate controls	
Protecting environment	Soil retention	
	Water retention	
Reuse water	Irrigation	
	Fishing	
	Washing	
	Swimming	Hygiene facilities

compensate the costs of sanitation. A brainstorm session with Enviu, business development firm specialised in the crowdsourcing, is organised to generate ideas for such services in communities. About twenty participants suggested many ideas for the wastewater services shown in Table 7.1. The services are mainly related to space use, such as playgrounds, parks and gardens. The unconventional ones are religious or ceremonial areas, recharging spots with solar energy, hygiene and washing, filters on roofs for gardening and climate control. Not many risks are expected if people are well informed about the non-potable effluent after treatment. The services based on wastewater add value and can generate income but cannot meet all effluent quality criteria in high-income countries unless septic tank technology is much improved.

7.4 Conclusions

The question if costly lock-in in a dominant technology network can be resolved giving chance to sustainable innovations is discussed with regard to sanitation. Sanitation is a pivotal service for social health and environmental qualities, but it is presently unavailable to nearly one-third of the global population. This service remains deficient in the next decades if the presently dominant network involving long transport through sewage and adding technologies for wastewater treatment prevails. This elongation of the sanitation chain effectively reduces risks connected with human excrements but it needs large investments entailing high capital costs unaffordable for many communities. The costs of additional sanitation increase faster than the technology change. The elongation network is locked-in, which is strengthened by the maxi–min policymaking, and it is aiming to maximise risk reduction at minimal cost.

Three alternatives can be induced based on different policy criteria. One alternative is focused on separating wastewater streams at pollution sources with the aim to reduce water loss and reuse matter in wastewater. This reduces costs at the sources, entailing cheaper sewage and wastewater treatment but it needs behavioural changes and non-sanitation supplies of architects and social science. Policies that prioritise the loss prevention can invoke this technology network. The second alternative is the distributed constructed wetlands, which treat wastewater under soil surface nearby pollution sources. This reduces the costs of capital and operations but it needs much space and incorporation in the spatial planning. The policy that aims to reduce investments can induce the constructed wetlands. The value-adding services based on use of wastewater can add value which may outbalance the costs and generate net benefits, but this sanitation method cannot attain all stringent standards and needs design and economic know-how in addition to the sanitation know-how. The policy aiming at value creation can generate such services. Each technology network has pros and cons; combinations can improve quality but increase the costs. Having the costly lock-in created by the policies throughout the last 150 years has created a need for diversification in sanitation. Given the health and environmental standards, the policy that enables communities to decide based on various criteria invokes the alternative sanitation technology networks. This diversification policy can resolve the lock-in entailing sanitation services accessible to all people.

References

Arceivala, S. J., & Asolekar, S. R. (2008). *Wastewater treatment, pollution control and reuse*. New Delhi: Tata-McGraw Hill Publishing.

Beder, S. (1990). Early environmentalists and the battle against sewers in Sydney. *Royal Australian Historical Society Journal, 76*(1), 27–44.

Burian, S. J., Nix, S. J., Pitt, R. E., & Durrans, S. R. (2000). Urban wastewater management in the United States: Past, present, and future. *Journal of Urban Technology, 7*(3), 33–62.

Cooper, P. F. (2001). Historical aspects of wastewater treatment. In P. Lens, G. Zeeman, & G. Lettinga (Eds.), *Decentralised sanitation and reuse: Concepts, systems and implementation* (1st ed.). London: IWA Publishing.

EPA. (1999). *Constructed wetlands treatment of municipal wastewaters*. Cincinnati: United States Environment Protection, mimeo.

Haller, L., Hutton, G., & Bartram, J. (2007). Estimating the costs and health benefits of water and sanitation improvements at global level. *Journal of Water and Health, 5*(4), 467–480.

Hutton G., & Batram, J. (2013). Global costs of attaining the Millennium Development Goal for water supply and sanitation. *Bulletin of the World Health Organization, 86*, 2–3. Visited 10-4-2013 http://www.who.int/bulletin/volumes/86/1/07-046045/en/

Hutton, G., & Haller, L. (2004). *Evaluation of the costs and benefits of water and sanitation improvements at the global level*. Geneva: World Health Organization, mimeo.

Hydrik. (1998). *Design manual, constructed wetlands and aquatic plant*. US Environmental Protection Agency, EPA/625/1-88/022, mimeo.

Krozer, Y. (2006). Projecting costs of emission reduction. In G. Meijer, et al. (Eds.), *Heterodox views on economics and the economy of the global society* (Mansholt publication series – Vol. 1, pp. 255–268). Wageningen: Wageningen Academic Publishers.

Krozer, Y., Hophmayer-Tokich, S., van Meerendonk, H., Tijsma, S., & Vos, E. (2010). Innovations in the water chain – Experiences in the Netherlands. *Journal of Cleaner Production, 18*(5), 439–446.

Krozer, Y., Franco-Garcia, M.-L., & Micallef, D. (2013a). Interaction management in environmental policy. *Management Research Review, 36*(12), 1210–1219.

Krozer, Y., Krozer, M., & Vos, T. (2013b, May 22–24). *Toward a beneficial sanitation.* International Workshop, Advances in Cleaner Production, Sao Paulo.

Kyambadde, J., Kansiimea, F., Gumaeliusb, L., & Dalhammar, G. (2004). A comparative study of Cyperus papyrus and Miscanthidium violaceum-based constructed wetlands for wastewater treatment in a tropical climate. *Water Research, 38*, 475–485.

March, J. G., & Olsen, J. P. (1984). The new institutionalism. Organisational factors in political life. *The American Political Science Review, 78*(3), 734–749.

Nelson, N., Tredwell, R., Czech, A., Depuy, G., Suraja, M., & Cattin, F. (2006, July 10–12). *Worldwide applications of wastewater gardens and ecoscaping: Decentralised systems which transform sewage from problem to productive, sustainable resource.* Paper for international conference on decentralised water and wastewater systems, Environmental Technology Centre, Murdoch University, Fremantle, WA.

Nelson, M., Cattin, F., Tredwell, R., Depuy, G. Suraja, M., & Czech, A. (2013). *Why there are no better systems than constructed wetlands to treat sewage water: Advantages, issues and challenges.* http://www.wastewatergardens.com/pdf/2007,SMALLWATspain.pdf. Visited on 7-4-2013.

Rockefeller, A. (1997). Civilization & sludge: Notes on the history of the management of human excreta. *Current World Leaders, 39*(6), 99–113.

Tousignant, E. (1999). *Guidance manual for the design construction and operations of constructed wetlands for rural applications in Ontario.* Stantec Consulting Ltd R&TT, Alfred College (University of Guelph) and South Nation Conservation, Canada.

UNDP. (2006). Human development report. Beyond scarcity: Power, poverty and the global water crisis. United Nations Development Programme, New York, mimeo.

UN-HABITAT. (2008). *Constructed wetlands manual.* Nairobi: UN-HABITAT.

WHO. (2012). *Global costs and benefits of global costs and benefits of drinking-water supply and sanitation interventions to reach the MDG target and universal coverage.* Geneva: World Health Organisation.

WHO/UNICEF. (2012). *Progress on drinking water and sanitation: 2012 update.* New York: United Nations Plaza.

Chapter 8
Alternatives for Commuting

Can project developers reduce traffic congestions? The traffic congestion is largely caused by commuting. However, better transport hardly reduces congestion because the commuting growth is an external effect of dislocating offices and residential areas, which is amplified by policy support of real estate development. The social cost of this external effect exceeds 16 % of the annual average wage. Life cycle costs of three innovative office systems, which embrace office work and commuting, are compared to the present office system: concentration in mega offices and distribution in local offices and home offices. As shown in this chapter, these innovative office systems save 15–28 % of the present office costs, reduce congestions and improve environmental qualities. The savings are USD 3.5–8.5 billion per million employees in the high income countries. Sensitivity analyses show the lower social costs of the local for nearly all urban conditions. Project developers can allocate funds into the ICT-based distributed offices, link offices with hospitality and suchlike, and policies induce this allocation when they abolish support of the wasteful infrastructure and internalise the external effects in the land prices for real estate development. The distributed office systems save USD 3.5–8.5 billion per million employees in the high income countries.

8.1 Real Estate Cycles

If the daily life rhythm in the last two centuries can be symbolised by walking through a factory gate, the present time can be by the daily traffic congestion when people commute to offices. Worldwide, employees drive cars during work days between 6 AM and 9 AM from home to offices and between 16 PM and 19 PM back home. This pattern is observed 5 days a week in all high-income and emerging economies; commuters in the low-income countries often use public transport. The high concentration of travellers during a few hours a day unavoidably causes congestion with negative external effects. There are productivity losses in distribution. Welfare losses are among others stress, car accidents and pollution. Social networks are disrupted when traffic moves through communities, and nature is degraded when landscapes are fragmented by infrastructure. The traffic congestion is often considered a transportation problem to be resolved within the domain of transport through more roads and better management of traffic flows. This viewpoint is attractive for the interests related to road constructions and traffic operations, but it does not help to reduce the congestion. Herewith, it is argued that commuting is caused by the present office work system and innovations in this system enable to mitigate congestions and reduce social costs of office work.

The office work involves nearly half of all labour in the high-income countries. Much of this work is data processing and interactions based on the information and communication technologies (ICT). This is largely individual work, but people are compounded in the office areas on a distance from the residential areas. The growing commuting mileage is due to the larger distances between the office and residential areas. It can be an individual choice to live far away from work, but spread of the work and home sites is not a personal choice. This spread, called urban sprawl, is driven by dislocation of the offices and residential housing. This dislocation is caused by cyclic imbalances of demands and supplies on the real estate markets. The cyclic imbalances occur because it takes time before constructions of housing can meet growing demands with associated high prices, and when the demands saturate along with the price drop, the constructions cannot stop immediately because of the past commitments (such imbalances are called the pork cycles). The high demands peak every 20–30 years, but the peaks for the office and residential housing do not evolve synchronous. People follow work with a time lag.

Since service work expanded in 1960s, demand for offices has grown. The offices until 1960s concentrated in the city centres grew on the city edges. This shift's purpose was to avoid the soaring prices in the centres. The residential housing followed with a time lag and by the time many residents of the city centres moved to the cheaper houses on the city edges, economic crisis, the oil crisis of 1980s, hit the real estate market. Offices downtown became vacant entailing prices to drop. Offices in the centres became payable. New offices moved in and new residents followed this work location. The economic boom of 1990s pushed up the office prices in the cities, and many offices moved out of the cities to suburbs. The construction of the suburban residential areas followed. The financial crisis in 2008 caused again high office vacancy and lower prices. Each cycle has increased distances between the offices and residential areas because the locations are shifted when vacant offices pauperize; a redevelopment of a pauperized area may take many decades. The urban sprawl enlarges the cars' mileage because public transport is sufficiently to reach scattered office areas and the transport by bike and walking is too slow for long distances. Commuting with the associated traffic congestion should be considered an external effect of the real estate market.

The cycles on the real estate market are largely spontaneous processes but policies can amplify or dim it. The present policies usually amplify this process because they support the real estate development. This support covers among others tax exemptions for loans on housing, entitlements for new real estate locations and funding infrastructure for the locations, as well as obstructing liabilities for the land use, social disruption, environmental degradation and other negative external effects of the real estate markets. In addition, the regional and local policies usually attract offices and residents because they are the main source of municipal income, and this income usually determines the politicians' wages and clergy employment. The local policymaking has direct interest in the real estate development. Policies could dim the real estate cycles through the internalisation of the external effects, but liabilities and compensations for harms are rare and deficient.

8.2 Mitigation of Congestion

The commuting mileage grows faster than the total travel mileage. For example, in the Netherlands between 1985 and 2010, the annual travels including international ones have grown from 120 to 170 km per person, but the share of commuting has grown from 21 to 34 %, mostly by car (nationale mobiliteit monitor and CBS woon-werk statistiek). The average commuting distance has increased from 12 to 18 km one way. By personal car, it was on average 22 km in 180 days a year in 2010.

Commuting causes high social costs. A commuter's direct user cost is the transport cost, which is mainly car use. When the car depreciation, gasoline, insurance, maintenance and taxes are included, the annual commuting transport costs by car are typically about €5150 per person (BerekenHet.nl, autokosten) although the depreciation and maintenance cost cannot be attributed solely to commuting. The indirect user costs are the travel costs, which cover infrastructure and travel time. The costs of constructing an extra motorway lane are €4–6.5 (USD 5–8.5) million per kilometre, and if they are capitalised including maintenance, these annual costs are about €0.5–0.8 (USD 0.65–1.04) million a year. It is equivalent of 0.6–1.4 euro-cent (0.8–1.8 dollar cent) per kilometre per traveller in the heavy traffic areas (Archer and Glaister 2006).

The cost of road use per person is a minimum of €70 (USD 90) a year usually paid through taxes. The travel time is typically 50 min one way excluding incidental traffic jams, repairs and suchlike during 180 days a year. The commuting time is an opportunity cost. If to assume 50 % average wage, which is about €16 (USD 21) per hour as being the opportunity cost, the travel costs are €4800 (USD 6240) a year. In addition, there are non-user costs. These costs are not covered from the private expenditures but paid through collective arrangements for accidents, noise, waste, breakthrough in communities and so on and to be paid in future for climate change, air pollution, fragmented landscape and others. The social costs of the collective arrangements are estimated to be a minimum of €470 (USD 611) per person a year in Europe (Essen et al. 2011; Korzhenevych et al. 2014). The bequest and existence costs of the transport systems are not found though these can be high because transport can undermine social inclusion of communities, amenities of the city life and threaten biodiversity. The total social costs of a typical car commuter in the Netherlands approach €10,490 (USD 13,637) a year, which is about 22 % of the average salaries. The imbalances on the real estate markets cause annually nearly

€42 (USD 54) billion costs to four million Dutch commuters by car. Not many individuals can avoid these costs because the office work is usually a collective arrangement though regulations are introduced to allow work at home unless the office work is required. Many people do not sense these costs when the commuting costs are compensated or exempted from taxes, but if people have to pay, these costs have large impact on the property value (Tse and Chan 2003).

Many scholars advocate improvements of mobility. The demand-side policies can put tax on fuel and traffic, foster selective car use and pooling, restrict parking, regulate speed and flow, inform people and so on. The supply-side policies can enlarge infrastructure, improve public transport, limit traffic in areas, discourage car ownership and improve traffic management, such as peak sharing. Technology policies can foster fuel saving cars, telematics for routing and new logistic systems. Physical planning can regulate land use to foster compact cities (Nijkamp 1994). In addition, services can be shared with drop-off points and bus transits, and liveability can increase through pedestrian and environmental and low-traffic zones (Goldman and Gorham 2006). Shifts from individual car travels to services and mobility management can be generated through tolls and road pricing (Nykvist and Whitemarsh 2008). Many nontechnical innovations are also advocated (Hyard 2013), such as pricing (parking, public transport fares and taxes), regulations (access control, parking fleet, informing, carpooling, dial and ride, staggered activity time, teleworking) and infrastructure (park and ride, pedestrian and cycling zones, public transport, ramp metering). Many suggestions are introduced to some extent in cities, but they do not outweigh the mileage growth of commuting.

Changes in the commuting patterns can also reduce congestion. Swapping the work and home locations and optimal routes planning would prevent 87 % of the commuting mileage in the United States (Hamilton 1982). This is criticised for not accounting trade-offs, such as different commuting destinations of couples and multifunctional travels to various locations, because people make shopping, take kids from school and so on (White 1977, 1988), and for neglecting imperfections on route because of road works, accidents and suchlike hinders (Small and Song 1992). Nevertheless, flexibility in the location and timing of the office work can reduce congestion. The ICT would enable office work on distance called telework. It would also shorten the commuting distance because it optimises travels through better access of locations, diversity of transport modes and so on, and it would make commuting more attractive through multitasking and suchlike (Wee and Chorus 2009). The ICT is used for the optimisation of commuting but the telework grows slowly. The ICT has hardly invoked telework in the rural areas despite large investments in the tele-networks in some regions (Grimes 2000). It is also a minor factor in substituting the office work for telework in the urban areas where the high age and education are much more important factors for this substitution (Graaff 2004; Graaff and Rietveld 2007). In result, the share of telework in the total employment in Europe varies from close to nil in Italy to 2.6 % in the United Kingdom for the full-time telework and from 2.3 % in Italy to 14 % in Denmark for the part-time telework. The growth of telework in Europe is estimated to be about 5 % a year (Vitola 2012), which is hardly above the growth rate of commuting and therefore cannot reduce congestions. The telework is lower in the United States where the travel distances are larger (Telecommuting 2014). Various social barriers for the telework are mentioned, such as social relations at work and appreciation of being away from home. The telework could also put pressures on wages and it needs space at homes (Riesen 1997). Present policies also facilitate commuting rather than the teleworking.

8.3 Alternative Offices

The social costs of a few innovative office systems are compared to the present system using life cycle costing (based on Enk et al. 1999). The life cycle costing is a label for various methods aiming to estimate costs throughout the life cycle of a system, which means in production, distribution, consumption and disposal during the system lifetime. This method is often used for the cost assessments of durable products and capital goods, in particular the infrastructural works. In essence, it is cost calculation per step in life cycle for a lifetime of goods. Details can be found in several handbooks (Dhillon 1989; Fabrycki and Blanchard 1991; Booz-Allen and Hamilton 1999).

The system, herewith, is the office work with commuting. Solely, the physical inputs are used, and the labour and capital inputs are considered constant. The output performances of all systems are assumed equal. The data is based on the Netherlands, which is typical for densely populated high-income countries. Mainly statistical data is used. Four systems are compared. The present office is the reference. This covers offices of various sizes in city centres and on city edges. One innovative system is the mega office. The assumptions are high concentration of office work in a suburban area and commuting a few kilometres more one way than to the present office mainly by public transport. Another innovative system is the local office. The assumptions are a distributed system of small offices with about 50 workplaces for rent on time basis and meeting points on a larger distance, all accessible from homes by the slow transport, i.e. walking, biking and suchlike. The last alternative is the home office with extra office space at home but without commuting. These office systems can be found in many countries on small scale. The costs are assessed in euro per employee and in total for one million employees, equivalent to an area of five million people. Four office inputs are considered: accommodation space, materials, equipment and commuting. These inputs largely determine the offices' performance given labour and capital.

The accommodation space is assessed with the statistics of offices and verified with a study on the space use of offices in the United States and the Netherlands (Bisseling 1998). Per person, the present office covers on average 35 m^2 gross for work inside the building and 20 % extra outside the building for car parking, travels and meetings at a third of the office square metre price. For the mega office, the statistically observed largest scale office category is assumed. The mega office is 74 % of the present office space. A smaller space outside the building is assumed proportional to the lower car transport in all transport. The local offices are nearly 66 % of the present office proportionally to the external meetings and absenteeism observed in the statistics on the office work. This lower space is due to rent per time unit. For the home offices the statistical smallest office per employee is assumed, which is 20 m^2 gross extra at the present office price.

Main materials are energy and paper. Energy covers heating and electricity for equipment, air conditioning and lighting. Statistical data on energy in the financial services is used. Compared to the present offices, the mega offices use more energy, and the local offices and home offices use less energy, which is based on very large, small and very small offices, respectively. The unit price of energy use is equal in all cases (though the scale is relevant for the energy prices). The paper use is based on the German banks' uses (Rauberger 1998), which are about 175 kg per employee a year for the present office. The mega office is assumed to use 20 % more paper for internal communication, the local offices 7 % less and the home offices 40 % less paper, which is based on various types of offices in Germany. In all cases, the same unit costs of paper are assumed.

The office equipment reflects the situation in the Dutch technological institute (TNO), an organisation with a few thousand employees (use of this information is highly appreciated). It covers furniture, ICT equipment, ICT experts and networks. All alternatives are assumed to use similar furniture, but the local offices use 30 % less furniture per employee due to time-sharing. All are supposed to be fully ICT equipped, among others docking station, desktop monitor, laptop, laser printer, server, fax, repro copier, fast modem and networks. Scale effects in the purchases and use of ICT are neglected. For the local offices, 30 % less equipment per employee is assumed due to the time-sharing, which is a conservative assumption compared to the present ICT utilisation rate, but faster depreciation of the ICT is assumed because of the intensive use. Hence, the ICT unit costs in the local offices are higher. The home offices ICT equipment is similar to the present office. The costs of the ICT services in the present system are estimated based on the average work and travel time of an ICT expert at the institute. An ICT expert serves 38 work units. This is also assumed for the mega office. In case of the local offices, only 30 work units per ICT expert are served because they must travel more and care about more equipment, and for the home office, additional 15 % costs are assumed to cover even more travels. A high hourly salary of the ICT experts is taken. The network costs are included in all alternatives, but in the local and home offices, these costs are assumed to be higher than in the present and mega offices.

The commuting mileage is assessed with the transport statistics. The train and car costs per kilometre are data delivered by the public transport enterprises and automobile associations. The unit cost of bicycles is a guess. The cost of walking is neglected. The average speed is assessed per transport mode for a few typical commuting routes. It is 50 min one way trip by car at 24 km per hour average speed. The trains are faster but include walking. Biking is slower but the travel distance is shorter. The transportation costs exclude car depreciation because cars are used for various activities. For the present offices, the statistical travels in the Netherlands are used: 55 % car; 10 % public transport subdivided into train, bus and metro; 30 %

bicycles and 5 % pedestrian. Most countries have a higher car and public transport share because bicycles are hardly used for commuting. The mega offices are assumed to be reached with public transport by 80 % of the employees and 20 % by car. The local offices are reached by biking and walking, each one by 50 % of the employees. The home offices do not need commuting. Per trip, the average transport costs vary from €3 (USD 4) for the present offices up to €7 (USD 9) for the mega office because of the larger distances, no slow transport and public transport. Reaching the local office is low cost because of biking and walking. The travel costs are travel time multiplied by the 50 % average hour wage, i.e. €16 (USD 21). Basic data is in Appendix. All data in the Appendix are based on Enk et al. (1999).

8.4 Life Cycle Costs of Offices

Table 8.1 shows the life cycle costs per employee. The office systems are shown vertically. Horizontally, the total costs of the office work and commuting are shown with the main cost factors as percentage of the totals. The present office work costs about €23,400 (USD 30,420) per employee per year. The innovative ones are cheaper: the mega office 12 %, the local office 22 % and home office 28 %. The most costly factor in all cases is the accommodation, which is 43 % of the total costs but 37 % of the mega office costs due to the higher density. The costs of energy and paper cover about 15 % of the present office costs, but they are higher in all alternatives up to 23 % of the home office costs. Paper use is the highest cost factor except for the home office where little paper use is assumed. The equipment costs, in particular the ICT services, increase from 16 % of the present office costs up to 27 % of the local offices and even 34 % of the home office costs. All these refer to the office work. Compared to the present offices, the work in mega offices is 15 % cheaper due to higher density; and the local and home offices are 7 and 5 % cheaper because the lower accommodation costs outweigh the costs of additional ICT

Table 8.1 Life cycle costs of the office systems

Main cost factors	Present office	Mega office	Local office	Home office
Total costs in €	23,384	19,892	18,150	16,840
Cost saving in %	0	−12	−22	−28
Space	44 %	37 %	44 %	43 %
Materials	15 %	18 %	21 %	23 %
Equipment	16 %	19 %	27 %	34 %
Subtotal	76 %	75 %	91 %	100 %
Work related in €	17,736	14,909	16,582	16,840
Transport	5 %	12 %	0 %	0 %
Travel	19 %	13 %	9 %	0 %
Subtotal	24 %	25 %	9 %	0 %
Commuting related in €	5649	4983	1568	-

services. The commuting costs are transport costs (means) and travel costs (time loss). Due to cheap biking and walking in the Netherlands, the transport costs are 5 % of the life cycle costs. This share can be much higher in the thinly populated countries and in the countries without facilities for the slow transport. The travel costs are 19 % of the life cycle costs. The mega office may cause higher transport costs but lower travel costs because of more public transport and shorter travel time. The commuting costs to local offices and home offices are low and nil, respectively.

Table 8.2 shows the result for one million employees in a year. Vertically, the office alternatives are shown and horizontally the main cost factors. The results are underpinned with four sensitivity analyses. Firstly, it is assumed that the office space is 50 % cheaper, for instance, in the suburban areas or economic peripheries. Secondly, the travel costs are neglected, which implies that people do not care or enjoy the commuting time. Thirdly, extra ICT costs are assumed to cover situations of high demand for experts. The last one is a combination of all of these, which is hypothetical because office work on periphery, joy of commuting and scarce expertise are rare combinations. Cost savings in all cases are possible. The costs of the present office system would be €23 (USD 30) billion per million employees per year. The potential cost savings would be €3.5 (USD 4.6) billion for the mega office system, €5.2 (USD 6.8) billion for the local office system and €6.5 (USD 8.5) billion for the home office system. These calculations are based on the minimum commuting costs because car depreciation, waiting time in incidental traffic jams and the costs of external effects are excluded. The sensitivity analyses confirm the cost

Table 8.2 Life cycle costs for one million employees with sensitivity analyses

	Present office	Mega office	Local office	Home office
In € billion a year				
Space use	10.3	7.4	7.9	7.2
Material costs	1.2	1.3	0.9	0.9
ICT costs	6.2	6.2	7.7	8.7
Transport costs	1.2	2.5	0.0	0.0
Travel costs	4.5	2.5	1.6	0.0
Total	23.4	19.9	18.2	16.8
Savings	0.0	3.5	5.2	6.5
Indexed				
High-cost space	100	85	78	72
Low-cost space	100	89	78	73
No travel costs	100	92	88	89
High ICT wage	100	84	73	78
No travel costs, low space and ICT costs	100	99	86	112

savings. Only if the office space is cheap, people enjoy commuting, the ICT is costly, and the home office system is more costly than the present one. The local office system is the most cost-effective alternative.

8.5 Conclusions

Congestions impede productivity in distribution and cause stress, air pollution, accidents and so on. The main cause of congestion is commuting. Commuting in turn is caused by dislocation of offices and residential areas, which is driven by imbalances on the real estate markets. Commuting with the traffic congestion associated to it should be considered an external effect of the real estate markets. Policies often amplify this effect when they support the real estate development. The social costs of this external effect, however, are high. In the high-income country as the Netherlands, the direct social costs of commuting approach are 22 % of the annual average salary, but the individual employees and employers cannot avoid these costs because alternatives need changes of the office systems. When the life cycle costs of the present office system are compared to the three alternatives – concentrated in mega offices, distributed in local offices and dispersed in home office – it is found that congestions would vanish and costs would be saved if innovative office systems are adopted. The savings per million employees could be as high as €3.5 (USD 4.6) billion for the concentrated office system up to €6.5 (USD 6.8) billion for the distributed office systems. The mega office systems have scale advantages in accommodation and public transport. The local office systems reduce commuting to nearly nil, which outweighs the additional ICT costs needed to facilitate such distributed office system. The home office systems have nil commuting, which outweighs the costs of the extra office space at home and high ICT costs. The outcome is robust for the low-cost space, costless travel time and high ICT costs. Shifts towards the innovative office systems are socially beneficial, and the project development aiming at the ICT-based office systems is profitable. The distributed office system emerges, which is reflected in cafes, restaurants and clubs populated by young people with the mobile ICT equipment. Project developers, therefore, can embark on this trend and diversify locations, tool and methods for the office work tuned to various office work specialisations. These options are socially beneficial, add value to the real estate and revive communities, and they may generate innovations because they enable more and diverse social interactions and novel work methods. Policies can enhance the innovative office systems when they internalise the social costs of the real estate markets in the land use prices through permits, fees on housing areas, liabilities and compensations for the harmful activities. This policy reduces congestion with high social benefits.

Appendix

Table 1 Space and costs of office working place (Enk et al. 1999)

Working place (€/m²)	Present office	Mega office	Local office	Home office
Work place gross	35	26	23	20
Parking	23	13	0	0
Others	11	8	5	4
Total	69	47	28	24
Work place (230)	7976	5991	5324	4608
Parking (69)	1563	884	0	0
Others (69)	791	536	319	276
Total costs	10,329	7411	5643	4885
Home, m² (184)	46	46	46	46
Public space, m²	97	97	130	130
Public space (69)			2304	2304
Office and living	10,329	7411	7947	7189
Index	100	72	77	70
Additional costs		−2919	−2382	−3140

Table 2 Costs of office equipment

Equipment	Present office	Mega office	Local office	Home office
Depreciation	1157	1157	1258	1009
Extra network centres	0	0	441	441
Extra network district	0	0	91	274
Interest	208	208	248	203
Labour costs	3842	3842	4866	5763
Software	393	393	393	393
Copier lease costs	11	11	11	11
Furniture	636	636	424	636
Total	6248	6248	7733	8729
Index	100 %	100 %	124 %	140 %
Additional costs			1486	2481

Table 3 Costs of office materials

	Present office	Mega office	Local office	Home office
Paper use kg/year				
Printing paper	109	130	101	72
Packing	39	46	36	26
Newspaper and books	16	19	15	10
Sanitary paper	5	6	5	3
Others	7	8	7	5
Total use	175	209	163	116
Costs (€4,6/kg)	806	963	753	535
Index	100	119	93	66
Energy use				
Additional costs		157	−54	−272
Lighting kWh	1468	1299	812	1468
Air cond. electric kWh	991	932	485	1109
Air cond. gas m^3/year	915	644	230	1066
Office equipment	622	523	361	684
Total energy use, GJ/year	53,2	39,5	16,5	60,8
Lighting in €	101	89	56	101
Air cond. electric	68	64	33	76
Air cond. gas	141	99	35	164
Office equipment	43	36	25	47
Total in €	352	288	149	388
Index	100	82	42	110
Additional costs		−18	−58	10

(*) m3 = GJ 46/1000, kWh 3,6/1000

Table 4 Data on commuting to offices

	Present office	Mega office	Local office	Home office
Cars				
Time minutes	51	21	0	0
Distance (km)	21	44	0	0
Travel cost (€)	16	7	0	0
Transport cost (€)	5	10	0	0
Percent commuters	55 %	20 %	0 %	0 %
Public transport				
Time minutes	44	44	0	0
Distance (km)	44	44	0	0
Travel costs (€)	7	7	0	0
Transport costs (€)	6	6	0	0
Percent commuters	10 %	80 %	0 %	0 %

(continued)

Table 4 (continued)

	Present office	Mega office	Local office	Home office
Bicycle				
Time minutes	26	0	9	0
Distance (km)	4	0	2	0
Travel costs (€)	8	0	3	0
Transport costs (€)	0	0	0	0
Percent commuters	31 %	0 %	50 %	0 %
Pedestrians				
Time minutes	16	0	18	0
Distance (km)	1	0	2	0
Travel costs (€)	5	0	6	0
Transport costs (€)	0	0	0	0
Percent commuters	4 %	0 %	50 %	0 %
Total	100 %	100 %	100 %	0 %
Costs (€)				
Travel cost per trip	12	7	4	0
Travel costs per year	4468	2512	1555	0
Transport costs per trip	3	7	0	0
Transport costs per year	1181	2471	12	0
Total per trip	16	14	4	0
Total annual (*)	5649	4983	1568	0
Index	100	88	28	0
Additional costs		−666	−4081	−5649

Mobility in office alternatives, 180 days * 2 trips in present, mega and local offices

References

Archer, C., & Glaister, S. (2006). *Investing in roads, pricing, costs and new capacity*. London: Imperial College, mimeo.
Bisseling, T. (1998). Alternatieve werkplekken: de eerste harde cijfers. *FACIO, 12*, 9–12.
Booz-Allen, & Hamilton. (1999). *California life cycle benefit-cost analysis model*. California Department of Transportation, Sacramento, mimeo.
de Graaff, T. (2004). *On the substitution and complementarity between telework and travel: A review and application*. Free University, Amsterdam, mimeo.
de Graaff, T., & Rietveld, P. (2007). Substitution between out-of home and at home work: The role if ICT and commuting costs. *Transportation Research Part A, 41*, 142–160.
Dhillon, B. S. (1989). *Life-cycle costing* (1st ed.). Amsterdam: OPA.
Enk van der, R., ten Houten, M., & Krozer, Y. (1999). Kosten van kantoorconcepten, TNO Delft, mimeo.
Fabrycki, W. S., & Blanchard, B. S. (1991). *Life-cycle cost and economic analysis* (1st ed.). Englewood Cliffs: Prentice Hall Inc.
Goldman, T., & Gorham, R. (2006). Sustainable urban transport: Four innovative directions. *Technology in Society, 26*, 261–273.

References

Grimes, S. (2000). Rural areas in information society: Diminishing distance or increasing learning capacity. *Journal of Rural Studies, 16,* 13–21.

Hamilton, B. W. (1982). Wasteful commuting. *Journal of Political Economy, 90,* 1035–1053.

Hyard, A. (2013). Non-technological innovations for sustainable transport. *Technological Forecasting & Social Change, 80,* 1375–1386.

Korzhenevych, A., Denhe, N., Bröcker, J., Holtkamp, M., Meier, H., Gibson, G., Varma, A., & Cox, V. (2014). *Update of the handbook on the external costs of transport.* London: Ricardo-AEA.

Nijkamp, P. (1994). Roads toward environmentally sustainable transport. *Transport Research A, 28A*(4), 261–271.

Nykvist, B., & Whitemarsh, L. (2008). A multi-level analysis of sustainable mobility transition, Niche development in UK and Sweden. *Technological Forecasting and Social Change, 75,* 1373–1387.

Rauberger, R. (1998). *Environmental reporting: The Vfu-case of benchmarking with environmental indicators in the banking sector, continuity, credibility and comparability, international expert seminar in Eze, France, 13–16 June, 1998.* Augsburg: Institut fur Management und Umwelt Mimeo.

Riesen, van F. (1997). *Ruim baan door telewerken.* Nederlandse Geografische Studies, Utrecht/Delft, dissertation.

Small, K. A., & Song, S. (1992). "Wastefull" commuting: A resolution. *Journal of Political Economy, 100*(4), 888–898 (Reprint of University of California).

Telecommuting. (2014). http://globalworkplaceanalytics.com/telecommuting-statistics. Visited 20-10-2014.

Tse, C. Y., & Chan, A. W. H. (2003). Estimating the commuting cost and commuting time property price gradients. *Regional Science and Urban Economics, 33*(6), 745–767.

van Essen, H., Schroten, A., Otten, M., Sutter, D., Schreyer, C., Zandonella, R., Maibach, M., & Doll, C. (2011). *External costs of transport in Europe.* Delft: CE.

van Wee, B., & Chorus, C. (2009). *Accessibility and ICT: A review of literature, conceptual model and research agenda.* Delft Technical University, Paper BIVEC-GIBET day, Brussels.

Vitola, A. (2012). *Overview on the European policies on telework.* Alborg: Intereg IVc.

White, M. (1977). A model of residential location choice and commuting by men and women workers. *Journal of Regional Science, 17*(1), 41–52.

White, M. (1988). *Urban commuting journeys are not "Wasteful".* University of Michigan, Ann Arb, mimeo.

Chapter 9
Ethical Consumers and Producers

Do consumers induce ethical consumption or companies seeking credibility for sales? This issue is discussed using assessments of the consumers' behaviour and experiences with life cycle management of firms considered leaders in corporate social responsibility. Consumers state high preference for the ethical consumption but do not reveal it in purchases given product prices. This gap is often explained by the inconsistent consumers' behaviour and deficient purchasing conditions, but supplies do not allow for sound assessments of functional qualities and ethical attributes compounded in products. Credible suppliers are highly rewarded with price markups and cost savings because can accrue a market share at low cost, save costs in supply chain and deliver value-adding products. A model on the consumers' and suppliers' deliberations between the functional qualities and ethical attributes, given prices, shows that the additional consumer demands have little positive influence on the ethical consumption compared to the innovating suppliers. Fostering sustainable innovations contributes more to ethical consumption than moralising about consumers' behaviour.

9.1 Consumers' Will

A scientific debate also popular on parties is if people have free will or they are instruments of a god, gene, brain or any other. An economic debate, herewith, is about the consumers' freedom in buying and using products and services of various qualities called the consumers' sovereignty. The issue is whether consumers determine qualities of supplies through their choice of products, or suppliers determine the consumers' choice through supplied alternatives. It is a non-trivial issue regarding nearly USD 1000 billion yearly global expenditures for external marketing aiming to attract consumers (Statista 2014).

The conventional viewpoint is that the consumer choice determines the supplied qualities, given product prices. The neoclassic, mainstream economists assume that consumers have a set of quality preferences called utility. Given their utility, consumers choose among product qualities. Axiomatic is that consumers aim to maximise their utility in transactions with producers that aim to maximise their profits. Having these aims, resources are spontaneously allocated via market negotiations towards the Pareto optimum when all are better off without anyone worse off. A deficient allocation is caused by market imperfections, for instance, because policies hinder competition and firms monopolies markets. This viewpoint that the consumers' demands determine supplied qualities given prices is reflected in policies, for instance, in the European Union 'Unfair Contract Terms' Directive (93/13/EEC) that enables to dissolve contracts in case of misinformation. This viewpoint is criticised. The consumers would not choose but be driven by greed to gain more than others and compare wealth rather than maximise utilities. Mass consumption would emulate the rich, leisure class whose utilities are mainly ceremonial, and the consumers' lifestyles would be expressions of such ceremonies (Veblen (1892) 2013; Trigg 2001). The utility concept would be tautology; consumers attach high value to a product because they prefer it, which obscures differences between the subjective value and the market value of products although these differ a lot (Robinson 1962); for example, a pet has high value to someone but be a cost on markets. Firms and consumers would not maximise profits and utilities because these are unattainable having many factors to deliberate in decision making. Market decisions would be driven by aspirations aiming at the satisfactory results given conditions and objectives of organisations in which decisions take place (Simon 1991). This opinion is also expressed in policies, such as the policy controls of product quality with the aim to provide reliable information (Berg 2011). The alterglobalist critiques on consumption go steps further. Corporations would create images that suggest consumers' choice, but markets are monopolised with socially and environmentally harmful products (Klein 2000). Marketing that celebrates brands as lifestyles would be only tools of sales (Fuat Firat and Dholakia 2006). Policies would individualise consumers' decisions but impede the organised consumers' actions (Hilton 2009). Despite many critiques, the consumer sovereignty remains the theoretical stronghold as shown in review on history of consumption (Sassatelli 2007).

Herewith, the ethical consumption is addressed, that is, purchases and uses of products and services with attributes of social responsibilities, among these being the attributes of environmental qualities. In the environmental economic train of thought, also referred to as the 'new economics', ethical values would drive the empathic and cooperative consumers' behaviour rather than competition and selfishness as argued in the conventional argumentation. Consumers would satisfy their personal and societal needs rather than maximise private utility and pursue the non-marketable goods rather than monetised services (Seyfang 2009). A comprehensive review on the consumers' opinions and behaviour suggests that most consumers do not consider ethics in their decisions but they sense responsibility for impacts of their consumption on other people and the environment. The ethical consumption is about the consumers' social responsibility and the consumer deliberation about functional qualities and ethical attributes of products and services (Devinney et al.

2010). This train of thought about high consumers' capabilities is sympathetic but could be wishful thinking. It is discussed whether the consumers should be supported in deliberations about functional qualities and ethical attributes, or the suppliers encouraged for credible supplies, given products prices and scarce budgets.

The functional qualities indicate performance in use, for instance, 'strong materials', 'flexible service' and others. The ethical attributes address the social responsibilities, for example, 'safe', 'natural' and so on. Each product may have several functionalities and attributes that comply with customers' criteria. Labels indicate the suppliers' credibility with respect to the qualities and attributes, and many suppliers show several labels certificated by an external organisation. Labels about the functional qualities show information about production (e.g. location) and product (e.g. handling manual). The producers in many countries are obliged to deliver this information, which is controlled by authorities. An additional functional quality is generally associated with value for money. Many products have labels with information about the ethical attributes, such as religious prescriptions, working conditions and environmental impacts. This information also addresses production (e.g. fair trade and ecological) and products (e.g. energy-saving and natural ingredients). An additional label suggests compliance with more social interests. The producers can get a label certified by authority, for instance, the Blaue Engel label in Germany and White Swan label in Scandinavia paved way for the ecological label in Europe. There are also private labels for the ethical attributes. The functional qualities can match ethical attributes (e.g. durability and energy saving), but trade-offs are usual (e.g. flexible supplies or ecological ones). The match usually adds credibility and market value. The credible products have a price premium. The trade-off can undermine the credibility and the value. Much information about the product qualities and ethical attributes can be found though reliability can improve. The issue is about decision making rather than information.

9.2 Ethical Consumption

The ethical consumption in the high-income countries during the first half of the last century was largely related to membership of churches, trade unions, cooperatives and social organisations. These memberships were rewarding because they entitled people to get housing, education, care, leisure and so on. From the 1960s onwards, such social arrangements gradually dissolved and evolved into markets, which is enhanced through policies aiming at more market negotiations than direct regulations. The memberships evolved in 1980s into the individual responsibilities of consumers without tangible rewards but advocacy of voluntary responsibilities for social and environmental issues (Nichols and Opal 2005; Litter 2009). Pools suggest that these responsibilities are embraced. In the 1990s, nearly one-third of all consumers in the high-income countries have stated preference for the products with ethical attributes (Hoevenagel et al. 1996). Mid-1990s in Germany, 35 % of consumers have stated that they buy ethically labelled products from time to time, and 16–35 % have stated that they are willing to pay a premium price for such products (Heiskanen and Pantzar 1997). The share of the German consumers that demand the ethically labelled products has increased from 19 % in 1989 to 46 % 10 years later (Imkamp 2000). Studies on the households' opinions in the high-income countries confirm high willingness to pay for environmental protection and environmentally sound products and services (OECD 2008, 2014). The stated preferences for the ethical consumption remained high during the last few decades. This, however, is not revealed in the consumer expenditures. As underpinned in Sect. 3.3, the ethical consumers' purchases do not exceed 2 % of all consumers' expenditures in the high-income countries; it is lower in the low-income countries.

A lot is hypothesised about the gap between the high stated preference and the low revealed preference. A popular explanation is compensation for affluence: people would be willing to consume ethically when they can afford to soothe bad conscious about wealth. This argumentation refers to the theory on satisfaction of needs expressed as pyramid of Maslow (Maslow 1943). People would satisfy their needs sequentially starting with a broad basis of the physiological needs (e.g. food), then smaller safety needs (e.g. health), belonging (e.g. family), esteem (e.g. leisure) and on top the smallest need of self-actualization (e.g. morals). The ethical consumption would be on top of this pyramid as a way of showing responsibilities. However, the Maslow theory is disputable. Another observation is that people living in the subsistence economies spend most time on leisure, ceremonies and social interactions in the community, time being an approximation of money in the non-market economies. The needs expressed on the pyramid are fulfilled in parallel to each other. The activities that fulfil social responsibilities are even more time-consuming than the activities aiming at the basic needs (Sahlin (1971) 2011). The high preference for the ethical attributes could reflect social conventions rather than individual choices; it is observed in behavioural experiments with consumers (Mazar and Zhong 2010; Vringer et al. 2013). Various practical barriers for the ethical consumption are also pinpointed. There are product-related barriers: the labelled products are costly, and quality is perceived inferior to alternatives, labels obscure and so on. The purchasing conditions also cause impediments: deliberations need much time, locations with labelled products are inaccessible, traditions and brand loyalties impede changes, adults learn poorly and suchlike. Better policies are also advocated: better information, certifications, infrastructure, pricing of external effects and others (Osbaldiston and Schott 2012; Thøgersen et al. 2012; OECD 2014). All these can be relevant in particular cases, but the general assumption that individual consumers can take decisions about the ethical consumption can be disputed and revival of the social arrangements for the ethical consumptions advocated (Power and Mont 2010).

It could be that the stated preferences indicate the consumers' demands for the ethical consumption and the gap with the revealed preference in purchases is due to the deficient supply of the ethical attributes. This deficiency means inferior product qualities compared to alternatives, given the product prices. It is plausible that assessing qualities compounded by subsequent suppliers in a life cycle of product requires high consumer skills and complex technologies and that sound assessments show many trade-offs within and between the functional qualities and ethical attributes.

The consumer deliberation of product qualities is complex and needs 'consumers' technologies' (Lancaster 1966a, b). Having such trade-offs, two unsolvable problems in the individual consumers decision making emerge. Firstly, it is a high risk and laborious decision making. Suppose consumers deliberate two functional qualities (e.g. 'durable' and 'lightweight') and two ethical attributes (e.g. 'eco-look' and 'fancy appeal'), each with 50 % chance of match with any other one. Then the chance to make right decision is only 0.5^3, or 12.5 %. If a product covers several qualities and attributes to be considered, as it is usually found on markets, the chance of right decision is very low. Similarly, the costs of deliberation or the transaction costs with suppliers are high, and social conventions are necessary to reduce these costs of decision making. The intractable dilemmas in the consumers' decision making can be avoided only when there is good match between and within the qualities and attributes in supplies. The match is perceived as the suppliers' credibility. The matching is also supported by the specialised services at a cost, such as energy service companies, car sharing and lease firms (Halme et al. 2004). Secondly, the ethical consumption refers to altruistic behaviour because the price of the ethical consumption is usually higher compared to alternatives; it is because supply of an additional attribute given quality is generally a cost. Altruism, herewith, is considered in the conventional evolutionary sense meaning the choice of giving fitness for another at a cost of own fitness when fitness is the ability to produce offsprings. Given any initial set of interests, the altruistic behaviour is an optimal strategy if this choice can be made repetitively and it is repetitively rewarded by the interests (Simon 2006). Inter alia, it is an application of the 'tit-for-tat' strategy developed in behavioural economics (Weimann 1990). However, the rewarding social mechanisms for the ethical consumption are largely lost during the decades of individualisation, except in communities with strict religion, high morality and ideological prescriptions. A few percent of ethical consumption found in the total consumer expenditures reflects, statistically, the scale of such communities.

9.3 Ethical Supplies

The supplies that match product qualities with ethical attributes are perceived to be credible and worth a premium price. They are highly rewarded. For example, the Bodyshop with personal hygiene products and campaigns for animal welfare was taken over by the L'Oreal for 910 million euro in 2006, and Ben and Jerry's with ice cream and ecological activism is taken over by the Unilever for 250 million euro in 2012. Many producers supply ethically labelled products, retailers sell them, and corporate social responsibility is used as a marketing tool for the ethical consumption although also as a cover up of deficiencies (Preuss 2001; Ramus and Montiel 2005). Suppliers that match functional qualities and ethical attributes in the life cycle of products save costs and gain income on markets (Krozer 2008). Three types of corporate results can be achieved as illustrated with three cases of life cycle management for the consumer markets. The case on herbs and spices shows how integration of ethical attributes in the conventional products enables to enter new markets. The textile case shows how this integration can generate cost savings in supply chain even irrespective of the customers' demands. The television case shows how matching ethical attributes can generate value added in consumption. The firms' goals, activities, costs and benefits and innovations, as well as the trade-off and match, are briefly presented. These firms became international, environmental leaders due to these innovations.

In the first case, a European wholesale leader in spices and herbs on the food and medical markets pursued organic supplies from the East Africa as an ethical attribute meaning production without chemicals. The food market is large by the materials volume but low priced, and the medical market of the herbal medicine and homoeopathy is a small volume market but high priced. The life cycle covers cultivation, processing, packaging at wholesale in Europe, distribution and retail sales mainly in Europe. The regular supplies are compared to the organic supplies for foods and medical uses. The costs are indexed based on the regular cultivation; all costs are averaged for several products. The consumer price of the regular herbs and spices for foods is nearly a 100 times the cultivation price (24 times cost increase to the wholesale followed by four times increase to the consumer prices). The costs of organic cultivation are three times higher than the regular one because of lower yields. The consumer price is nearly twice higher, which makes the supplies for foods uneconomic. The costs of organic cultivation for the medicinal purposes are nearly seven times higher, but the consumer price of medicinal products is about 260 times the cultivation prices, which makes this market attractive for the supplies of ethical attributes. The firm has integrated ethical attributes in supplies for the medical markets. On the food market, there is a trade-off between the functional quality and ethical attribute, but there is match on the medical market.

The textile case is based on supply chain of a large retailer. When the organic textile was introduced in boutiques, this company followed, but customers did not buy, and the textile was discounted at a loss. The company's attention turned to the supply chain from India, aiming to eliminate hazardous labour conditions and risky additives in textiles and prevent losses of packaging and clothes during transport. A cotton men's shirt is used as a test case. The life cycle covers cultivation, garment production and confection in India, then lengthy transport by trucks and ships for export and short transport for upgrading of clothes and sales in retail in Europe. The main hazards are pesticide use in cultivation, toxic additives and waste water in production, unsafe work in the confection and wasteful and energy-intensive distribution. The consumer price is 40–60 times the cost of cotton cultivation (4–5 times increase for confection, 3–4 times increase in trade and a similar increase in retail). The ecological cultivation is 2–3 times costlier than the regular one; the eco-dying is also much costlier. There is trade-off because of these; they do not contribute sufficiently to the value added in sales, and risks in sales are high because the most costly factor is the discounting of the textile when unsold because it goes out of fashion. The firm focused on better work conditions and materials and energy savings in the supply chain entailing cost savings equivalent of a few percent of sale prices. The integration of ethical attributes in the assortment failed, but it was successful in the supply chain without tangible consumer demands.

The television case is about an electrotechnical producer. The German law on recycling of electronic goods triggered this firm to anticipate similar regulations in other countries (similar is introduced by many countries with a time lag). The main environmental impacts and costs of strict environmental policies are assessed in the life cycle of a television. The component production causes hazards of heavy metals and solvents. Low impacts are in distribution. Consumption causes waste of packaging and repairs and pollution at electricity production. The discharge on landfills generates hazardous waste. These steps cover 41 %, 38 %, 20 % and 1 % of the life cycle costs, respectively. Strict environmental policies would mainly increase the consumption costs, which are mainly caused by pollution controls at electricity production. The recycling legislation would have a negligible effect on the life cycle cost. The company has focused on the largest cost increase in the life cycle, i.e. energy use in consumption, and developed a prize winning energy-saving television. This has set trend for the international energy label on televisions. The recycling is outsourced at low extra cost. The ethical attributes caused trade-off with some functional qualities but triggered profitable innovations because of the added value to consumers. There is trade-off between ethical attributes and functional qualities in the supply chain but a match in the sales.

The cases illustrate that producers have motives for integration of the ethical attributes in products even without tangible consumer demand and in cases of trade-off with functional qualities. Regarding the consumers' stated preferences for ethical consumption, firms can generate a new market, save costs in supply chain or add value to the functional qualities when integrating ethical attributes in products.

9.4 Decision Making

A model is presented to analyse deliberations of consumers and producers that face a trade-off between the ethical attributes and product qualities. The mainstream economic approach is used in which consumers and supplies allocate resources between the functional qualities and ethical attributes (adapted after Krozer 2004). Possible interactions between the consumers and suppliers are shown. Assumed is an assortment of products with several qualities and attributes. Adding any of these is a cost and matching becomes more costly as qualities or attributes are added. Firstly, an individual consumer and producer are considered. Then, consumers and producers on a market are discussed.

The model is presented in Fig. 9.1. The X-axis represents the additional functional qualities and the Y-axis the additional ethical attributes. Each one shows the increasing costs of consumption. The consumer's expenditures are allocated between these two values, which is represented by the lines D_1 and D_2 which are the lower and higher demand for the ethical attributes. The supplier deliveries are shown with the lines S_1 and S_2 having the lowest, l, and the highest, h, boundary of the

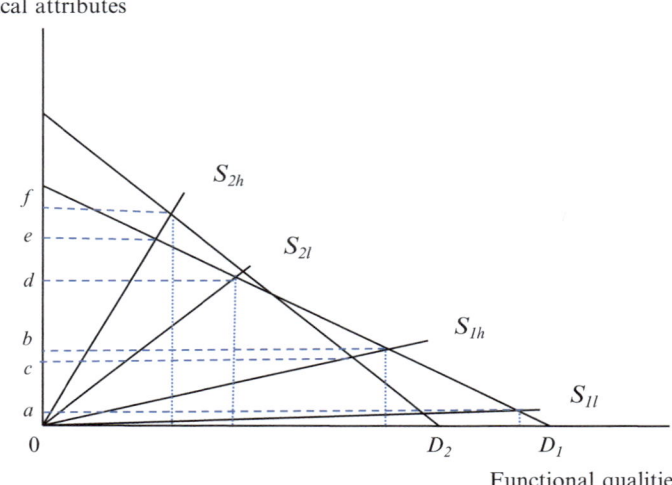

Fig. 9.1 Consumers and producers decision making about functional qualities and ethical attributes

assortment. In the initial situation, the consumer demand, D_1, and the producer supply, $S_{1l} - S_{1h}$, deliver high functional qualities but low ethical attributes, $b - a$. An increase of the consumer demand for the ethical attributes is relative to the functional qualities, D_2, but unchanged supply has a low positive effect or a negative effect on the ethical attributes shown by $b - c$. The negative effects occur when some ethical products are taken out of the assortment $D_1 - D_2$ because of the perceived risk for sales of other products (so-called product cannibalism). An extra consumer's demand for the ethical attributes without supplier's innovation hardly invokes more ethical consumption and may even impede it. If the supplier innovates through integration of ethical attributes in the supplies, $S_{2l} - S_{2h}$, given the initial demands, D_1, the ethical attributes increase $d - e$ and even more, $e - f$, if the demand increases to D_2. A non-innovative producer impedes the ethical consumption even if ethical attributes are highly demanded, and an innovative producer reinforces the ethical consumption even at low demand.

A market situation with several demanding consumers and competing suppliers is more realistic and shows similar results. The assumption about demands for the ethical attributes relative to functional qualities is the demands' decrease as more attributes are added. The assumption about supplies is that an additional ethical attribute triggers others to add the attributes (irrespective trade-off with the functional qualities and demanded or not) until saturation with the ethical attributes and competition causes net firms' costs. Figure 9.2 shows the market situation. Similarly to the Fig. 9.1, the functional qualities are shown on the X-axis and ethical attributes on the Y-axis. The consumer's initial and increasing demands, D_i and D_{i+t}, respectively, and the initial and innovative supplies, S_i and S_{i+t}, respectively, are shown. Initially, the ethical attributes are added; insofar, they match the functional

9.5 Conclusion

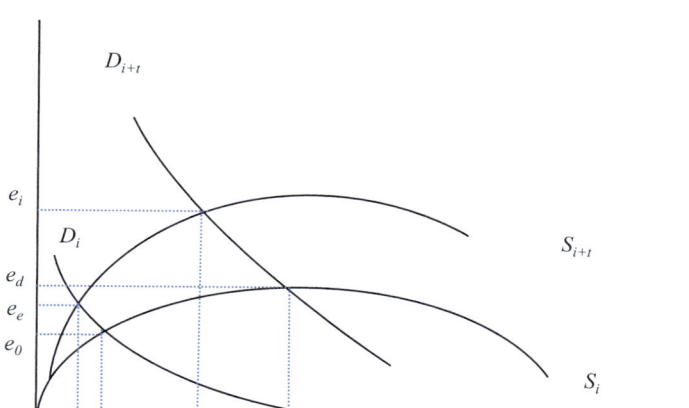

Fig. 9.2 Consumers and producers decision about functional qualities and ethical attributes

qualities, $0 - e_0$. If the suppliers do not innovate, the increasing demands for the ethical attributes add only little to ethical attributes, $e_0 - e_t$, compared to the addition of functional qualities, $q_0 - q_d$. If the suppliers do innovate and the consumer's demands do not increase, more additional ethical attributes are generated $e_0 - e_e$, but functional qualities can be degraded $q_0 - q_e$. If the suppliers innovate and the demands for ethical attributes increase, the additional ethical attributes may even surpass the additional functional qualities, $e_0 - e_i$ compared to $q_0 - q_i$.

9.5 Conclusion

The question is discussed whether the ethical consumption is mainly invoked by the consumer's demands for ethical attributes in products or it is due to the companies innovations when they pursue credibility entailing a premium price. The neoclassic economists, who underline the consumer sovereignty, would argue that the consumer's demand determines the supplied qualities. The high consumer's preferences for products with ethical attributes expressed in inquiries would support this viewpoint, but a small fraction of all consumers' expenditures allocated for such products would underpin the opposite perspective. It could be that the ethical consumption is nearly impossible for an individual consumer because disentangling trade-offs within and between functional qualities and ethical attributes in products is too complex and the altruistic ethical consumption unrewarded. The integration of ethical attributes in the producer's innovations can generate profits in the life cycle of products. Cases of corporations show profitable innovation when this integration enables entry on new markets, invokes cost savings in supply chain and adds value to functional qualities. In all cases, the consumers' preferences for the ethical attributes were indecisive for the innovations or even misleading. A model in which firms supply an assortment with functional qualities and ethical attributes and consumers deliberate between these underpins this train of thought. More consumer demands for the ethical attributes hardly increase the supplies and may even reduce it if the supplier perceives this demand a risk for other products. The suppliers' innovations aiming at more ethical attributes have a strong positive influence on the ethical consumption. Given the stated preferences, the ethical consumption can grow if suppliers innovate. The ethical consumption saturates when an overload of the ethical attributes causes net costs on markets; more consumers demands help only to postpone this saturation. Instead of moralising about an individual consumer's behaviour, consumer organisations and policies should encourage the ethical innovators.

References

Berg, A. (2011). No roadmaps but tools: Analysing pioneering national programs for sustainable production and consumption. *Journal of Consumer Policy, 34*, 9–23.
Devinney, T. M., Auger, P., & Eckhardt, G. M. (2010). *The myth of the ethical consumer* (1st ed.). Cambridge, UK: Cambridge University Press.
Fuat Firat, A., & Dholakia, N. (2006). Theoretical and philosophical implications of postmodern debates: Some challenges to modern marketing. *Marketing Theory, 6*(2), 123–162.
Halme, M., Jasch, C., & Scharp, M. (2004). Sustainable home services. Toward household services that enhance, ecological, social and economic sustainability. *Ecological Economics, 51*, 125–138.
Heiskanen, E., & Pantzar, M. (1997). Towards sustainable consumption: Two new perspectives. *Journal of Consumer Policy, 20*, 409–442.
Hilton, M. (2009). *Prosperity for all, consumer activism in the era of globalisation* (1st ed.). New York: Cornell University Press.
Hoevenagel, R., van Rijn, U., Steg, L., & de Wit, H. (1996). *Milieurelevant consumentengedrag, Ontwikkeling conceptueel model* (pp. 19–62). Rijswijk: Sociaal Cultureel Planbureau.
Imkamp, H. (2000). The interest of consumers in ecological product information is growing – Evidence from two German surveys. *Journal of Consumer Policy, 23*, 193–202.
Klein, N. (2000). *No logo* (1st ed.). London: Flamingo.
Krozer, Y. (2004). Social demands in the life-cycle management. *Greener Management International, 45*, 95–106.
Krozer, Y. (2008). Life cycle costing for innovations in product chains. *Journal of Cleaner Production, 16*, 310–321.
Lancaster, K. J. (1966a). A New approach to consumer theory. *Journal of Political Economy, 74*(2), 132–157.
Lancaster, K. J. (1966b). Change and innovation in technology of consumption. *American Economic Review, 56*(1/2), 14–23.
Litter, J. (2009). *Radical consumption, shopping for change in contemporary culture* (1st ed.). Berhshire: Open University Press.
Maslow, A. H. (1943). A theory of human motivation. *Psychological Review, 50*(4), 370–396.
Mazar, N., & Zhong, C.-B. (2010). Do green products make us better people. *Psychological Science, 21*, 494–498.

References

Nichols, A., & Opal, C. (2005). *Fair trade, market driven ethical consumption* (1st ed.). London: Sage.
OECD. (2008). *Household behaviour and the environment*. Paris: Reviewing the Evidence.
OECD. (2014). *Greening household behaviour*. http://www.oecd-ilibrary.org/environment/greening-household-behaviour_9789264214651-en. Visited 20-10-2014.
Osbaldiston, R., & Schott, J. P. (2012). Environmental sustainability and behavioral science, meta-analysis of pro-environmental behavior experiments. *Environment and Behavior, 44*, 257–299.
Power, K., & Mont, O. (2010). The role of formal and informal forces in shaping consumption and implications for sustainable society: Part II. *Sustainability, 2*, 2573–2592.
Preuss, L. (2001). In dirty chains? Purchasing and greener manufacturing. *Journal of Business Ethics, 34*(3–4), 345–359.
Ramus, C. A., & Montiel, I. (2005). When are corporate environmental policies a form of greenwashing? *Business & Society, 44*(4), 377–414.
Robinson, J. ((1962) 1964). *Economic philosophy* (1st ed.). Suffolk: Penguin.
Sahlin, M. ((1971) 2011). *Stone age economics* (1st Paperback ed.). New York: Routledge.
Sassatelli, R. (2007). *Consumer culture, history, theory, politics* (1st ed.). London: Sage.
Seyfang, G. (2009). *The new economics of sustainable consumption* (1st ed.). New York: Palgrave McMillan.
Simon, H. (1991). Organizations and market. *Journal of Economic Perspectives, 5*(2), 25–44.
Simon, H. (2006). Altruism in economic. *American Economic Review, 83*(2), 156–161.
Statista. (2014). Marketing spend worldwide 2009–2020 in http://www.statista.com/statistics/282197/global-marketing-spending/.
Thøgersen, J., Jørgensen, A.-K., & Sandager, S. (2012). Consumer decision making regarding a "green" everyday product. *Psychology & Marketing, 29*(4), 187–197.
Trigg, A. B. (2001). Veblen, Bourdieu, and conspicuous consumption. *Journal of Economic Issues, 35*(1), 99–115.
Veblen, T. (2013). *The theory of the leisure class*. http://www.gutenberg.org/files/833/833-h/833-h.htm. Visited, 23-8-2014.
Vringer, K., Vollebergh, H., van Soest, D., van der Heijden, E., & Dietz, F. (2013). *Dilemma's rond duurzame consumptive. Een onderzoek naar het draagvlak voor verduurzaming van consumptive*. Bilthoven: Planbureau voor de Leefomgeving.
Weimann, J. (1990). *Umwelt-ökonomik, eine theorieorientierte Einführung*. Berlin: Springer.

Chapter 10
Sustainable Investors and Innovators

Do public funders and private investors foster sustainable innovators? The global research and development expenditures grow faster than income, inventions are patented even faster and the venture capital growth rate for market introduction of innovations is even higher. All these activities involve risky private investments, but some costs and risks are covered with public funding. The private funding of sustainable innovations is high compared to all innovations. A higher policy support of sustainable innovations compared to all innovations could also be expected because these serve social priorities, but it is not in the Netherlands and possibly in other countries. When opinions of the interest groups in sustainable innovations about policy diverge, policymaking is at risk because interests can object it. Interviews with the sustainable innovators and sustainable investors underpin this argument. Co-operative models in financing reduce the innovators' and investors' risks.

10.1 Innovation Process

A global consensus is that policies should support innovations that generate large social benefit if this support does not distort markets. It is advocated because all innovation processes are risky. High costs must be made during several years before sufficient income is gained to cover all costs and make profit, which is possible when customers generate benefit, but this benefit is uncertain at the start of innovation processes. The sustainable innovators bear an additional risk. This risk is that they must make additional costs of developing technologies that foster environmental qualities, which can benefit many interests, but not many are ready to pay for these innovations if they do not gain individually. Regarding these social benefits but free-riding behaviour, one could expect more policy support of the sustainable innovators compared to all innovators. Why this expectation may not hold in many countries is illustrated based on experiences in the Netherlands where much public funding for innovations is provided and policy aiming at sustainable development is highly prioritised.

Innovators must spend throughout an innovation process. Expenditures are needed to generate and research good ideas and design inventions. These two phases are called research and development (R&D). Tests of the inventions, pilot production and scaling up for sales, which are called market introduction, need the so-called seed, start and expansion capital. Income can be generated only after these expenditures. Innovators, investors and the public suffer if failures are funded but successes are unpredictable at the moment of funding because incidents can occur in every phase of the innovation process. For example, the researchers can discover that wrong ideas are selected, designers can find deficiencies in research, tests can show that customers are dissatisfied with the invention, pilot can fail because of quality imperfections and production may not sell because demands have changed compared to the situation at the start. The chance of success increases in each phase but the investment as well. The investment multiplied by the chance of success indicates the investor risk. For example, if in each phase one million extra is expected to be needed for the successful launch of a new product, having 50 % chance of success per phase, then 114 million investments must be available: $1/0.5^5 + (1+1)/0.5^4 + (1+1+1)/0.5^3 + (1+1+1+1)/0.5^2 + (1+1+1+1+1)/0.5^1$. In reality, the investments and the chance of success fluctuate per phase and vary per innovation process. The estimates based on this demand-pull model suggest that huge expenditures are needed for any innovation. This could illustrate the expenditures in a large corporation and institutions; an anecdote is that the Pentagon disseminated this model in the 1950s to justify policy support of its huge research (Godin 2006; Godin and Lane 2013). Fortunately, innovating is usually a trial and error based on individual skills and copying ideas found in and knowledge spillover networks without such large investments as discussed in Chap. 1.

Innovators need risk-taking investments. Loans are difficult to get because the innovators' know-how has low security value. Innovators need equity to operate. Large corporations use internal funds if high returns are expected but the most risky phase, which is research, is usually outsourced to the public institutions. This way, the private costs and risks are reduced. The innovating start-up firms and small- and medium-size companies depend on the external private investors and public funding. The private investors in equity, called venture capital, also take risks, but being shareholders, they can influence decisions, and for this risk taking, they demand high interest; it can be as high as 25 % interest rate above the rate on loans. The public funding of innovators is a risk to tax payers, but it reduces the private investments, strengthens equity and signals credibility to the private investors, customers and policymakers. Due to the public support of innovators, the private financiers can take the risks of innovation investments, and if innovations can bring social benefits, there is reason for this support. This benefit justifies the policy support of sustainable innovations.

10.2 Investing in Innovations

The global expenditures on innovations grow despite the economic dips after the burst of the Internet bubble in 2001 and the financial crisis in 2008. The annual expenditure on research and development increased between 2000 and 2011 on average 0.4 % faster than the real global income growth of nearly 2 % annual average in that period (World Bank). Differences between countries, however, are large. The United States' average expenditures of € 943 (USD 1226) per person have decreased, as well as in Japan with similar expenditure per person. In the European Union, the average was € 428 (USD 556) and the expenditures increased. The increase is also observed in South Korea that has spent roughly half of the European Union per person, and a faster rate of increase is observed in Russia and China where the expenditures per person are lower than the South Korean ones. The countries' expenditures on research and development, when measured per person, seem to converge towards a level between the European Union and United States. The global results of these expenditures indicated by the number of patents have increased even faster by 4.8 % annual average during 2000–2012 (World Bank). It suggests that the research and development expenditures became more successful or the patents are easier to adapt. The market introduction evolved even faster. The average time span between the start of research and launching a new product on consumer markets is reduced by half every 10 years throughout the past decades (Ganguly 1999). This 50 % reduction in 10 years implies 7 % annual growth. The market entry of the business start-ups is largely financed by the venture capital. From 1995 to 2011 the venture capital in the United States has grown by an annual average of 15 %. The venture capital per person was averaged around € 63 (USD 82) per year in that period. During 2007–2011, i.e. after the financial crisis, the growth in the European Union was lower than in the United States, but the average of € 88 (USD 114) per person was higher (it is based on the Eurostat for the European Union and the PricewaterhouseCoopers participation statistics for the United States).

It is argued that more venture capital enables cheaper capital supply, which enhances research and development. The patents and patent citation functions of the venture capital growth in the United States industries during the 1980s would underpin this argument (Kortum and Lerner 2000). Similar effects are found with the cross-state data of the United States. Based on this finding, it is argued that an additional venture capital has a larger impact than an additional funding of research and development, but the argumentation is mainly semantic (Akin 2011). A lot is expected from the venture capital for the innovation policies in the high-income countries (Lerner 2010). The expectation could be too high. The positive effects of the venture capital on innovations in the United States could be during the boom periods when ample investment opportunities in innovations were found, in particular during the ICT boom in the 1990s, but the venture capital could be less important during the innovations' slowdown. Such positive links are not found in the European Union during the slump of 2007–2012. The cross-country correlations of venture capital with research and development expenditures per person (total and only private) and venture capital with patents per person show no positive results ($R^2=0.02$ and $R^2=0.05$, respectively). The research and development growth generates demands for venture capital, not the other way around.

Cost savings are also relevant in research and development. Shifting research and development to the lower income countries can reduce expenditures without performance losses. In the European Union, for example, the expenditures on research and development grow faster in the former communist countries with high technological abilities and low labour costs, such as the Czech Republic, Hungary and the Baltic States, than in the research-intensive countries, such as Austria, Germany and the Scandinavian countries. These countries catch up with their own resources and attract research and development from abroad. Shifting the private to public funding is also done. The private expenditures on research and development vary from 62 to 74 % of the total expenditures in the high-income countries, but they are shifted to the public ones in several countries, in particular in the United States. In addition, policy support of the private expenditures is partly unobserved statistically, for example, the public research for companies' priorities and tax support of the start-ups. A large part of all costs and risks of the innovating is covered by the public funding. More public support of innovation processes is advocated, but public support of the rent-seeking behaviour is not justifiable and can crowd out the cost-effective funding. Differences between countries also exist. For example, per person, the policy support of the private expenditures on research and development in the United States is much larger than in the European Union, and it grows faster.

10.3 Policy Support of Innovators

The private and public funding of all innovations is compared to the funding of sustainable innovations in the Netherlands. Sustainable innovations, herewith, are indicated by data on the energy and environment business. The private and public funding of research and development of all innovations and sustainable innovations is based on the national statistical data (CBS). The private funding of the market introduction of all innovations and sustainable innovations is based on the PricewaterhouseCoopers data. The public funding of the market introduction of all innovations is based on the annual report on innovations of the Ministry of Economic Affairs (EZ 2008), but the funding of the innovation dissemination is excluded. The public funding of the market introduction of sustainable innovations is studied by the largest employer association, FME-CWM. It identified 23 push and pull instruments, which means for technology developer and for the technology users, respectively (Grimbilas 2008). The data on the policy instruments cover only 2008 because more is unavailable. Table 10.1 shows the funding.

During the period 2001–2009, the annual average private funding of research and development was € 5100 (USD 6630); out of it, € 79 (USD 102) million was for sustainable innovations, which is 1.6 % of all private funding. The annual average

10.3 Policy Support of Innovators

Table 10.1 Public and private funding of research and development and market introduction in the Netherlands during 2001–2009 in € million annual average

€ million a year	All innovations	Sustainable one	Sustainable in all (%)
Research and development			
Private	5084	79	1.6
Public	3800	165	4.3
Sum	8884	244	2.7
Public in total (%)	43	68	
Market introduction			
Private	597	30	5.0
Public, only in 2008[a]	1380	64	4.6
Sum	1977	94	4.8
Public in total (%)	70	68	

CBS *Research and Development, Speurwerk en Ontwikkeling*. Private funding of seed, start-up and expansion (PriceWaterhouseCoopers 2002–2010). Public funding of the market introduction (EZ 2008 and Grimbilas 2008), with WBSO but not the diffusion instruments: SDE € 1.344 million (Grimbilas 2008), SenterNovem Annual report (2008) € 448 million, fiscal support € 425 million, Borgstellingregeling Scheepsnieuwbouw € 1000, € 263 million sustainable innovators (EIA, MIA and VAMIL)

[a]Private funding in 2008 was € 696 million; out of it, € 163 million was for sustainable innovations, i.e. 23 %

public funding of research and development was 3800 (USD 4940) million for all innovations; out of it, € 165 (USD 215) million was for sustainable innovations, which is 4.3 % of all public funding. The public funding of research and development on sustainable innovations is relatively larger than the private one, which is in line with the argumentation about high risks of the innovating for collective interests such as environmental qualities. The annual average private funding of the market introduction was € 597 (USD 776) million; out of it, € 30 (USD 39) million was for sustainable innovations. It was 5.0 % of the total amount. In 2008, the public funding of the market introduction was € 1380 (USD 1794) million; out of it, € 64 (USD 71) million was for sustainable innovations. This is 4.6 % of all public funding, which is well below 23 % of the private funding for sustainable innovations in that year. The mix private–public funding of sustainable innovations done by the regional development corporations was percent-wise as low as the public funding and decreased throughout 2001–2009, whereas the private investments increased by the average of 35 % a year.

The public funding of research and development of all innovations, which means the most risky phase of the innovation process, was lower than the private one, but the public funding of the market introduction was more than twice higher than the private one. This public funding is risk averse and encourages the rent-seeking behaviour because it is more attractive for the innovators to seek the policy funding than to compete for venture capital. The public funding can crowd out the private one. The opposite holds for sustainable innovations. The public funding of research and development is higher, and one of the market introductions is lower than the private funding. This policy supports the sustainable innovators' research and development, but it obstructs market introduction because it supports the rivals.

The number of innovators and funding per innovator based on that statistical data shows the chance of getting funding. All data refer to the period 2001–2009. The annual average number of innovators that received the private funding in the research and development phase was 6811; out of them, 106 were the sustainable innovators. These numbers were 210 and 22, respectively, in the market introduction phase. These numbers imply that on average, 3 % of all innovators received venture capital. A similar chance to get venture capital is found in many countries, experts say. The sustainable innovators had a much higher chance to get venture capital. On average about 21 % of them received private funding during that period. However, the average investment per sustainable innovator was about half of the € 2.8 (USD 3.7) million average investment per innovator. It should also be noted that sustainable innovators that got the private funding chance increased throughout the period 2001–2009, whereas the chance to get fund from the regional development corporations decreased. The private investors perceive the sustainable innovators more attractive and lower-risk investments than the public investors.

10.4 Investors' and Innovators' Interests

Compared to private investments, the policy support of research and development for sustainable innovations is larger than for all innovations, but the support for market introduction is smaller. Why would policy foster sustainable investors and punish sustainable innovators? How to comprehend this inconsistency? A strong anti-sustainability business lobby is an unconvincing argument because the largest employer association mentioned above supports the sustainable innovators. It could be an anomaly in the policy traditions or temporary mistakes of incapable politicians, but these are not plausible regarding the vested policy priority for sustainable development and high policy performance during several decades in the Netherlands. Instead of a policy deficiency, this inconsistency could be caused by the stakeholders involved in sustainable innovations. The explanation could be the opposing stakeholders' interests. Prime interest groups in the public funding of sustainable innovations are innovators and investors. If there is a consensus within an interest group but the groups' opinions differ or even collide – this situation is called the information asymmetry – the directly involved policymakers face the risk that one of the interest groups obstructs progress. If such obstruction is effective, the position of policymakers and ultimately the decision-making politicians is at stake. Given the information asymmetry, the policymakers' strategy to maximise effects of the lowest risk activity could prevail because it is the least risky policy, as discussed in Chap. 5. Much is theorised about the information asymmetry (reviewed in Barbaroux (2014)). The mainstream theory aims to reduce the information asymmetry, but the behavioural theory considers it as being an entrepreneurial resource.

10.4 Investors' and Innovators' Interests

The mainstream view on the information asymmetry is focused on market transactions. Customers who buy a product could be willing to pay a premium price for good products, but knowing that they can expect hidden deficiencies known only to suppliers, they tend to pay a price based on the average quality known on markets. This price, says the theory, is below the customers' willingness to pay for the high-quality products. The information asymmetry would deplore prices entailing lower quality (Akerlof 1970). In analogy, customers of innovators, being investors, would invest below the optimal capabilities of innovators because they suspect that innovators overestimate their capabilities or hide problems. Various methods are used to assess and counter the information asymmetry. One approach that embraces various methods is to characterise successful innovators based on their track records; it is an analogy to the selection of reliable sellers. The track record can refer to outputs, e.g. the innovators' revenues or consumer satisfaction, or to their inputs, e.g. the number of ideas or research and development costs in sales. Decisions, however, are biased when innovators without track records are assumed unsuccessful although the contrary is observed, among others, that the new firms and new managers in firms generate more successful innovations than the vested ones (Kaplan et al. 2009). Another method is focused on the innovation content, in analogy tests of products. Since investors are usually laymen about the innovator's subject, they often involve external experts. It does not resolve the problem because the experts are biased by their specialisation and the investors are biased in their choice of experts and interpretation of findings. It is observed that the mediocre proposals are often rewarded and the unconventional ones neglected because they are considered uncertain or unfeasible by the experts. Evaluation of evaluators is recommended (e.g. Perrin (2014)), but this enlarges administrative burden without problem solving when the evaluators are biased.

The contrary viewpoint is that the information asymmetry is not a deficiency but it is inherent to the entrepreneurial operations and necessary for discoveries. An essential entrepreneurial skill is in this view scanning and finding opportunities for discoveries due to the information asymmetry between innovators, investors, authorities, customers and other interests. Errors of some entrepreneurs are main resources of information for others (Kirzner 1997). Fostering the entrepreneurial skills in discovering opportunities is advocated through the knowledge spillover, public engagement, education and networking, awarding ideas and so on. Instead of creating bureaucracy to reduce uncertainties in the selection of successful innovators, which deplores innovation, enlarging capabilities for risk bearing is advocated (Shane 2000). The implication is that the high uncertainty provides innovative opportunities. "Innovation is precisely something that gains from uncertainty: and some people sit around and wait for uncertainty and using it as raw material, just as our ancestral hunters," writes Taleb (Taleb 2012). In this view, uncertainties are considered a resource because they generate a chance of success if only people are capable to take the chance and are enabled to act. The risk bearing reduces the rent-seeking behaviour, but entrepreneurs who can be ruthless cheaters can also undermine the valuable activities.

The information asymmetry also refers to sustainable innovations. Polls show that the sustainable innovators often assume that investors want to take over their know-how; they consider investors ignorant and expect more policy support. Vice versa, the investors argue that the innovator's proposals are of poor quality and they have low entrepreneurial capabilities, and they point out at the crowding-out effects of policy interventions (Oxford Research 2010). The inconsistent policy with respect sustainable innovations could be due to the information asymmetry between the innovators and investors that pursue sustainable development. The decision-making sustainable investors and innovators are interviewed to assess whether there is consensus about private and public funding. Fifteen investors are interviewed: seven equity and eight debt financiers. Fifteen innovators from various businesses are approached; 12 of them have co-operated. The selection of interviewees is illustrative for the case of information asymmetry, but it is not a representative sample. The same questions are presented to all, and minutes of the interviews are used after their approval. The innovators are also asked about the fate of their applications for the private investments and public funding by the regional and national authorities as they submitted nearly hundred applications. The findings are discussed in two workshops with about 40 persons each (one at the Ministry of Economic Affairs and one at the Association of Environmental Professionals). The presented results should be considered as validated. The results are summarised in Table 10.2. The opinions are scored: + 1 yes, 0 maybe and −1 no. A score above six points within a group indicates consensus within the group. A difference above six points between the groups indicates information asymmetry. The stakeholders' opinions per question and the innovators applications are presented.

Sustainability. The investors' opinions about the extraordinary risks of financing sustainable innovations are divided. Some consider this financing risky compared to all innovations, but others experience more social support of these innovations,

Table 10.2 Opinions about financing sustainable innovations

Consensus if ≥ |6| in group, asymmetry if difference ≥ |6|	Investors				Innovators			
	Yes +1	Nuance 0	No −1	Score total	Yes +1	Nuance 0	No −1	Score total
Is it exceptionally risky?	5	5	5	0	8	1	3	5
Is scaling up very difficult?	3	12	0	3	7	4	1	6
Are subsidies necessary?	7	5	3	4	5	2	5	0
Can banks finance innovations?	2	6	7	−5	1		10	−9
Is there sufficient risk-taking capital?	7	6	2	5	1	6	4	−3
Regulations instead of subsidies?	11	4	0	11	6	5	1	5
Are addition policy instruments needed?								
Quality certificate and label	12	3	0	12	5	2	5	0
Guarantee and suchlike	11	4	0	11	5	3	4	1
Support with know-how	13	2	0	13	9	1	2	7

10.4 Investors' and Innovators' Interests

which makes this financing easier. Many innovators pinpoint additional barriers for sustainable innovations such as deficient prices, disputable assessment methods and inconsistent policies, but there are also opposite views. There is no consensus and a minor asymmetry. *Scaling up.* The investors' opinions about the most risky phase vary. Many agree that the starters without a track record pose a high risk for investments, but others argue that the scaling up of sustainable innovations is risky. Most innovators have a similar opinion. A few investors and inventers find that every phase in the innovation process can pose difficulties. There is a consensus among innovators and no asymmetry. *Subsidies.* Various opinions about subsidies are expressed. Most investors agree on subsidies for research, development and demonstration but not for the subsequent phases, and a few investors advocate repayment of subsidies if profits are made, so-called revolving subsidies. Several innovators argue that subsidies are needed for all phases because sustainable innovations provide social benefits, but a number of them find that the private investments are more important. There is no consensus within the groups and no asymmetry. *Banks.* Most investors and nearly all innovators agree that banks are not suitable for financing sustainable innovations because risk of failure is high and that the innovating starters need equity, respectively. A few investors argue that the institutional investors should make more risk-taking investments. There are nearly consensus among investors and consensus among innovators and no asymmetry. *Risk capital.* Most investors perceive that there is no shortage of venture capital and advocate better marketing of the financial instruments. They argue that innovators should improve understanding of the financing issues. Most innovators perceive that the venture capital is scarce, and a few argue that more capital is found abroad. Many innovators find that the investors are myopic, demand too high interest and show little engagement. There is nearly consensus among investors and asymmetry between the groups. *Regulations or subsidies.* Most investors advocate more demanding regulations because policy support without larger markets reduces the chances of successful investments. Instead of subsidies, a few investors advocate fiscal instruments and regional financing, but others find this risky. Most innovators advocate more subsidies and worry that regulations undermine competitiveness. There are consensus among investors and asymmetry between the groups.

A few questions addressed policy instruments that can be added to the private and public funding. Three types of policy instruments are mentioned: (1) quality certificates and labels; (2) guarantees for prices, credits and suchlike; and (3) knowledge exchange and networks. *Certificates.* Nearly all investors would like a certification with strict sustainability criteria based on the transparent assessment methods. The innovators' opinions are divided. A few argue that the quality promotion is needed, but most find quality certificates and labels not needed because they obscure real quality and because criteria are biased. There are consensus among investors and asymmetry between the groups. *Guarantees.* Nearly all investors support more and better guarantees, e.g. for energy saving, cash flow guarantee during expansion, private–public guarantee fund and a long guarantee period for innovations, possibly in combination with subsidies for the initial phases of innovation processes. Several investors propose demand-driven policy instruments, such as engaging citizens as investors through crowdfunding and cooperatives, mixed public and private funding, assisting starters with local and regional funds, fostering innovation procurement with sustainability criteria and suchlike. Most innovators are critical about the guarantees and suchlike because of debatable criteria and bureaucracy. They prefer subsidies, possibly the revolving ones, or combined with the guarantees. There are consensus among the investors and asymmetry between the groups. *Knowledge*. Nearly all investors would like more knowledge support and exchange, for instance, about sustainability assessments. Many innovators also advocate knowledge support, but they prioritise networks with investors and policymakers, as well as administrative support for applications and transparent assessments. There are consensus among investors, nearly consensus between innovators and asymmetry between the interests.

There is reasonable consensus within one of the interest groups on most issues but barely any consensus between the groups except about inadequacy of the banks for innovations. There is information asymmetry between the interests with respect to the availability of risk capital, regulations and subsidies. The investors usually oppose more subsidies and advocate more demanding policies. The innovators generally wish more risk-taking support and subsidies instead of stricter policies. There is information asymmetry between the groups about the necessary additional instruments for sustainable innovations. The investors propose more quality certificates, labels and guarantees, as well as other instruments that enlarge markets of innovations which are largely dismissed by the innovators who prioritise innovation subsidies. The investors and innovators also ask for different kinds of knowledge.

The innovators are also asked about the results of their applications in the last 3 years. All their applications were about financial support of technology development and demonstration, none of them about management, marketing or financial affairs. The private funding is mainly venture capital. The public funding is divided into the regional and national funds. Ten out of twelve innovators applied for a private funding. Most innovators have financial partners, but only four always or regularly ask for a financial advice. Most innovators submit a few proposals to a few investors, and a few innovators spread many proposals to many investors. Out of 88 submitted applications to the private investors, 75 are rejected. The chance of success is 15 %, which is in line with the findings based on statistical observations. Reasons for the rejections according to the innovators are too innovative and risky, conflicting interests, too low equity, disagreement about scale and operational misfit. Nearly all rejected proposals are adapted and submitted to an authority. Five out of eleven innovators applied to the regional authorities in addition to the national funding. The main reason is spread of the risks. Out of 11 submitted applications, 5 are rejected, which means a 55 % chance of success. Reasons for rejections according to the innovators are different opinions about management, a mismatch with regulations and too small company. The rejected applications are resubmitted abroad and developed with the innovator's own cash. Ten out of eleven innovators applied for the national funding. Out of 33 submitted applications, 18 are rejected, which is 45 % chance of success. The reasons for these rejections in the opinions of the innovators are misunderstanding of the concept, a mismatch with regulations and assessed not feasible. Most rejected applications are reformulated and resubmitted to the private investors in the Netherlands and abroad. Most proposals get support after several applications. Seeking policy support of innovations does pay off.

10.5 Conclusions

A question mark is assigned to the assumption that the public funders and private investors foster sustainable innovators. The context for this issue is the faster pace of innovation processes. This is reflected in the growing expenditures on research and development, faster market introduction of novelties, shorter time to market of innovations and even faster-growing venture capital for innovations. The policy priorities for sustainable development could be reflected in larger support of research and development and market introduction of the sustainable innovators compared to all innovators because they contribute to the private and public income, as well as to environmental qualities. This is not found in the Netherlands for the period 2001–2009. The policy support of research and development on sustainable innovations as percentage of the total funding is small. It is percent-wise even smaller for the market introduction of sustainable innovations than for all innovations. The private investments in sustainable innovation are percent-wise to the total much larger than the policy funding. The competitors of sustainable innovators gain percent-wise to total much more public funding. The private investments in sustainable innovations increased, and the chance that the sustainable innovators get venture capital was high and increased, whereas the policy support stagnated and in some cases decreased. This policy impedes sustainable innovations. It is explained by the information asymmetry between the sustainable innovators and investors, because different opinions of the key interest groups can obstruct the policymaking. Indeed, the sustainable innovators and investors have different opinions. There is reasonable consensus within one of the groups on many issues but hardly any consensus between the groups, except about the inadequacy of banking for innovations. The interest groups have opposite opinions about three policy topics. The innovators argue that the venture capital is scarce, whereas the investors' opinion is that it is well available, which is confirmed with the statistical data. The innovators advocate more subsidies and higher chance of success, but the investors prefer more demanding policies and instruments that generate demand for sustainable innovations. The innovators observe the low policy supports of the market introduction when compared to all innovators and to the private investments. The investors advocate quality certification and guarantees for credits because these reduce investment risks, but these are largely opposed by the innovators because they confront higher costs. The opinions reflect their interests but can cause a risk-avoiding policy. In practice, the sustainable innovators apply for private and public funding with high success chance though the public funding is usually small compared to the private funding. The high chance of success is largely due to resubmission of the application. Persistence wins. The success rate depends as much on the number of resubmissions as on the quality of proposals. Co-operative models on financing sustainable innovations reduce the investors' and innovators' risks.

References

Akerlof, G. A. (1970). The market for "lemons": Quality uncertainty and market mechanism. *The Quarterly Journal of Economics, 84*(3), 488–500.

Akin, M. S. (2011). Does venture capital spur patenting? Evidence from state-level cross-sectional data for the United States. *Technology and Investment, 2,* 295–300.

Barbaroux, P. (2014). From market failures to market opportunities: Managing information under asymmetric information. *Journal of Innovation and Entrepreneurship, 3*(5), 1–15.

EZ, Ministerie van Economische Zaken. (2008). *Concurrend Ondernemingsklimaat.* Rijksbegroting.

Ganguly, A. (1999). *Business-driven research and development* (1st ed.). New York: MacMillan Business.

Godin, B. (2006). The linear model of innovation: The historical construction of an analytic problem. *Science, Technology & Human Values, 31*(6), 639–667.

Godin, B., & Lane, J. J. (2013). *"Pushes and pulls": The (hi)story of the demand pull model of innovations.* Montreal: Project on Intellectual History of Innovations, mimeo.

Grimbilas, P. (2008). *Duurzaamheid: De overheidsinstrumenten voor bedrijven.* Zoetermeer: FME-CWM, mimeo.

Kaplan, S. N., Sensoy, B. A., & Strömberg, P. (2009). Should investors bet on the jockey or the horse? Evidence from the evolution of the firms from early business plan to public companies. *The Journal of Finance, LXIV*(1), 75–115.

Kirzner, I. (1997). Entrepreneurial discovery and the competitive market process: An Austrian approach. *Journal of Economic Literature, 35*(1), 60–85.

Kortum, S., & Lerner, J. (2000). Assessing the contribution of venture capital to innovations. *The RAND Journal of Economics, 31*(4), 674–692.

Lerner, J. (2010). *Innovation, entrepreneurship and financial market cycles* (OECD nr. 2010/3). Paris: OECD.

Oxford Research. (2010). *Financing eco-innovation,* mimeo.

Perrin, B. (2014). *How to—And how not to—Evaluate innovation.* http://www.mande.co.uk/docs/perrin.htm. Visited 29 Aug 2014.

PriceWaterhouseCoopers. (2002–2010). *Ondernemend Vermogen 2002 t/m 2009.* Amsterdam: Nederlandse Vereniging van Participatiemaatschappijen.

SenterNovem. (2008). *Jaarverslag 2008.* Utrecht: Mimeo.

Shane, S. (2000). Prior knowledge and discovery of entrepreneurial opportunities. *Organization Science, 11*(4), 448–469.

Taleb, N. N. (2012). *Antifragile* (1st ed.). New York: Random House.

Chapter 11
Energy Services for Smart Grid

What are business opportunities and impediments for innovators in energy services? In the European Union, thousands of local energy initiatives emerge and many of them evolve into enterprises that pursue distributed energy systems called smart grid. They add value in energy consumption. The number of new enterprises on the energy markets increased by the annual average of 3600 firms along with average 23,000 additional jobs per year during the period 2008–2011. These newcomers grow mainly due to services in the renewable energy production, efficient solutions for the residential electricity and value-adding gas use in businesses. Price guarantees for renewable energy delivery to grid support these innovators. In addition to competition on the energy markets, these innovators must overcome impediments posed by energy taxation that supports large-scale energy consumers and subsidies for fossil fuel producers that are larger than subsidies for renewable energy. These policies obstruct energy efficiency and renewable energy. The distributed energy systems emerge despite annual €317 billion policy support of the vested, large-scale energy businesses.

11.1 Local Energy Initiatives

Thousands of local energy initiatives of citizen, farmers, small- and medium-size enterprises and non-governmental organisations emerge in Europe. Some scholars consider this to be a trend towards the democratisation of energy markets. The 'open', locally embedded decision-making processes would substitute the 'distant', decision making in private firms and the 'closed' institutional decision making in public utilities (Walker and Devine-Wright 2008). The community context of these initiatives would foster the open decision making. For example, case studies in Austria suggest that the community involvement generates renewable energy production that is tuned to the local demands (Ornetzeder and Rohracher 2006). Case studies in the United Kingdom show that know-how about the community-based energy systems generates social enterprises which serve social interests (Ison 2010; Seyfang and Haxeltine 2012). Case studies in the Netherlands also point out innovations on market niches, but minor system changes on the energy markets are expected and hybridisation of the local initiatives with the vested, large-scale energy business is advocated (Arentsen and Bellekom 2014). Risks are also underlined, in particular disparities between the organisers' expectations about community involvement and low participation (Rogers et al. 2008) and challenges related to professionalisation of organisations when the local energy initiatives grow beyond the community scale (Hielscher et al. 2011). Inquiries into the local energy initiatives in Italy, Romania and the Netherlands demonstrate hybridisation of the innovating firms with social and institutional involvement. The main success factors in the opinions of these initiatives are high motivation of participants, good relations in communities and sound business case; the main risk is poor economic performances (Boon 2012; Dragoman 2014).

From economic perspective, the local energy initiatives can be considered innovative entrants on energy markets. They add value due to functional qualities and ethical attributes of energy consumption, which address respectively the private interests and social responsibilities as discussed in Chap. 9. Figure 11.1 indicates these values of exemplary energy products. Some energy products are highly valued. Hence, the energy market can grow even if energy consumption decreases in energy terms. For example, electricity from grid is about ten times higher than energy equivalent of fuel, a unit stored electricity in a conventional battery about ten times the value of electricity equivalent from grid, and the hydrogen storage is a few times more valued than the conventional batteries. These values relate to convenience, flexibility, power storage and so on. The value also grows as carbon units per energy unit decrease mainly because of concerns about pollution, e.g. acidification in the past and climate change now. The value of decarbonisation fosters the renewable energy use. For instance, solar energy grew fast despite high costs during the last decades when matched social interests expressed in policies.

The local energy initiatives can be considered innovators in the sense that they serve customer-specific products and services. Such services are often perceived unattractive by the vested, large-scale energy producers that are focused on the bulky supplies. These innovators generate the distributed energy systems on the scale of buildings, districts and regions called smart grid. The term 'smart' refers to the advanced technologies and 'grid' to the electricity and heat technology networks. The energy systems embrace the value chains of fossil fuels that use coal, oil, gas and nuclear resources and renewable energy that uses biomass and waste, hydro, geothermal, solar and wind resources, as well as energy efficiency that covers energy saving and value addition to these uses. The communities that accommodate know-how, customers, capital and other resources create conditions for evolution of the civil society activities into the business start-ups entailing enterprises and business networks that deliver the customer-specific technologies and services often labelled as energy service companies. Such energy services expanded far beyond their community. For example, the Ecopower, energy cooperative in Belgium, captured about 4 % of the Belgian energy market with wind power and biomass for energy, and the wind- and solar-based firms in Navarra, Spain, operate on the international scale (Krozer 2012b). The possibilities of these innovators are discussed in the context of the European Union policy.

11.2 Energy Service Companies

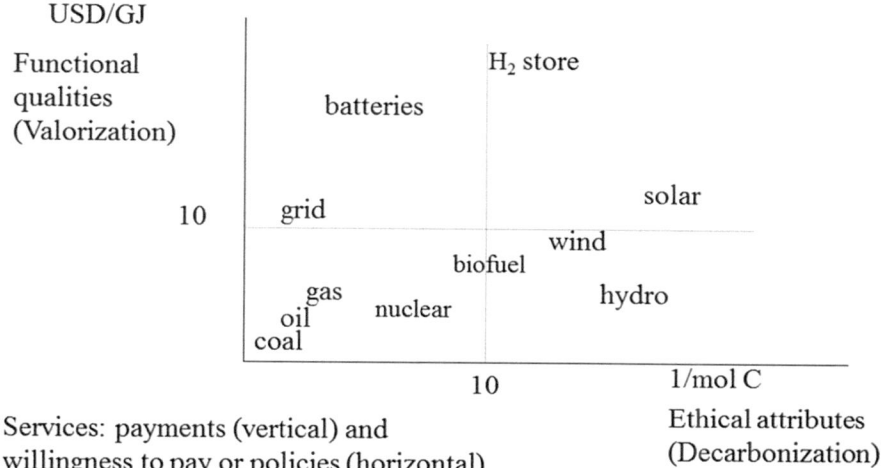

Fig. 11.1 Value of energy innovations

11.2 Energy Service Companies

The number of enterprises and jobs on the energy markets in the European Union during the period of economic crisis 2008–2011 is shown in Table 11.1. It shows all enterprises; enterprises in the information and communication technology business (ICT), which is considered a dynamic business; and in the electricity, gas and air conditioning business. These are compared.

The number of enterprises does not increase in the European Union despite 10 % annual birth rate because many firms collapse. The ICT, which covers about 10 % of all enterprises in the European Union, has generated a 13 % birth growth rate and somewhat lower death rate. As a result, the number of enterprises has grown by 3 % annual average. The number of energy enterprises is about 10 % of the ICT number, but they have grown by 24 % annual average due to 31 % annual birth and low death rate. The annual average growth of the energy business is about 3600 firms with 23,000 jobs, which is an impressive performance during the period of the economic crisis. The number of employees per firm decreases over the years, which can be a decay of the new firms or capitalisation per employer. The high turnover growth of these enterprises suggests the capitalisation. This implies that the energy service companies are not solely the consultancy firms but increasingly the technology-based firms. Appendix Table 1 shows the data on all countries in the European Union. The highest growth is in Czech Republic, France and Italy. These countries are also considered the European hot spots of the energy service companies and smart grid as defined in the European Union (Bertoldi et al. 2014).

Table 11.1 Number of enterprises in the European Union

	2008	2009	2010	2011	Growth (%)
All enterprises	23,398,388	23,485,779	23,464,546	23,690,350	0
Birth of new firms	2,289,532	2,272,681	2,292,839	2,301,626	0
% birth to all (%)	10	10	10	10	
Information and communication	914,401	938,763	963,606	1,002,459	3
Birth of new firms	116,640	117,517	124,747	128,736	3
% birth to all (%)	13	13	13	13	
Electric power, gas, air conditioning	59,015	73,879	94,874	113,178	24
Birth of new firms	10,099	15,363	22,863	20,937	31
% birth to all (%)	17	21	24	18	
Employees	1,242,176	1,273,412	1,300,322	1,309,951	2
Employees/enterprise	21	17	14	12	−18

Market opportunities for the energy service companies are estimated based on the volumes and prices of energy consumption. The estimates are done for the European Union of 27 countries, i.e. excluding Croatia, during 1995–2011 for the residential and business energy consumption on-site and in transport. The on-site consumption is based on the energy statistics. The consumption in transport is derived from the mileage per modality in the transport statistics on the assumption of constant energy consumption per mileage per modality. The shares of kilometre passengers for the residential transport and the kilometre-ton freight for the business transport in the total transport are calculated per modality based on 3-year average mileage 2010–2012 (assumed average travels are 1440 km for passenger ships, 3600 km for freight ships and 2270 km for flights and a kilometre passenger is 0.1 kilometre-ton). The annual average data on energy consumption is shown in Appendix Tables 2.a, 2.b, and 2.c: the residential and business on-site and in transport, the renewable energy consumption, the growth and the prices. The estimate for transport is done to tune with the on-site data; differences with the recent Eurostat estimates are not essential and not worth discussing in this context.

The results show that the total energy consumption has stabilised. About 1.1 billion ton oil equivalent a year is consumed, equivalent to 13,000 GWh, in which 26 % is residential, 43 % business, 25 % in passenger transport and 5 % in freight transport. The energy consumption per person is on average in the European Union, about 2.3 ton oil equivalent; it is 27 MWh. The Central and Eastern European countries consume less than 2 ton oil equivalent per person. Market growth in these countries is possible to match 3 ton oil equivalent in high energy consumption countries: Luxembourg, Finland and Sweden. The total on-site consumption in the European Union site has decreased on average by 0.2 % a year, and in transport the total energy consumption has increased by 1 % a year though a decrease should be expected in the near future due to the energy-efficient transport means.

11.2 Energy Service Companies

Table 11.2 Indicators of gas and electricity market in the European Union of 28 member states during 2004–2011

Annual averages (own calculations with the Eurostat data)	€/kWh	€ million	Annual increase (a)	
	Fuel to sale price (%)	Gross margin	Gross margin (%)	Sales prices (%)
Gas business	75	17,415	50	2
Gas, residential	23	12,312	8	5
Electricity business	26	62,222	4	5
Electricity, residential	7	43,474	2	1
Total		135,424		

(a) Annual average increase of the FOB gas price is 21 %, fossil fuel mix price is 10 %

The value growth despite saturation in energy terms in the European Union is in the distributed renewable energy systems and the customer-specific value-adding energy-efficiency technologies. The renewable energy consumption in the European Union has grown on average 6 % a year between 2002 and 2011 towards the 14 % share in its total energy consumption in 2011. The share of renewable energy is lower than the global 19 % due to high biomass consumption in the developing countries measured in 2011 (Sawin 2013), but a higher share is envisaged for 2020 in the European Union and progress is on track. The renewable energy consumption based on the domestic solar and wind production grows fast. It is mainly due to the distributed wind and solar farms, which are units ten to hundred times smaller than the conventional fossil fuel plants, and the residential production. The value-adding opportunities in energy efficiency are indicated by the gross margins, which are sales minus costs of the resource purchases. The gross margins are calculated for the gas and electricity consumption on-site. It is not elaborated for transport because the price data is statistically unavailable though energy efficiency is attractive due to the high prices and diversification of fuels. Table 11.2 shows four indicators for the residential and business gas and electricity consumption. Appendix Tables 3.a and 3.b shows the data of EU countries. The share of the fossil fuel price in the sale price in euros per kilowatt hour indicates the value addition: the smaller this share, the larger value added. The total gross margin indicates the market volume. The average annual margin growth shows the market growth and the sale price growth shows the price factor of the margin growth, which indicates price elasticity of demands. The value of the on-site electricity and gas services in the European Union is about € 135 (USD 175) billion a year (it is similar to the annual real estate sales and twice larger than the ICT sales). Note that this market volume is excluding purchases of resources.

The energy services for the residential electricity consumption generate the highest value added. This allows many high-tech services. These generate value-adding innovations, such as balancing of local grid, electricity storage, the electricity-efficient equipment, advanced lighting, low-voltage equipment, energy monitoring and management and so on. The lower value added per energy unit and a larger total market are services for the business electricity consumption. Innovations in the business electricity efficiency are attractive if they are large scale or disseminate fast because the lower value added impedes the distributed systems. These innovations can be, for example, advanced capacitors, direct for alternate current substitution, electricity storage and applications similar to the residential electricity use in offices, shops and suchlike. The residential and business electricity markets together cover about 78 % of the total value. Gas is mainly consumed for heat. The gas consumption is a smaller market by value because of low value added, but by the energy units in kWh, it is much larger than the electricity market. Innovations in energy saving are attractive on the business gas markets because margins on gas uses grow fast in Europe. Low-cost, large-scale services are needed with regard to the low margins, which is a major barrier for the energy service companies. The heat-saving equipment does not generate large cost savings unless equipment is of low cost and it matches the energy policy on mitigation of climate change. Even more attractive are services that add value to gas, such as transformation of heat to electricity and heat exchange. The residential heat use is nearly 20 % of all on-site energy use in the European Union, but only low-cost innovations are attractive because the heat loss has low value. Insulations and heat management particularly in old buildings are attractive. It is not per se because of the cost recovery due to saved heat but thanks to the energy labels for housing because a higher label enables a higher price of sales or rents.

The margins of such energy services have increased in all cases. The margins of the residential electricity also increased, but the increase is lower than the increase of the FOB fossil fuel prices. Herewith, the diversification of energy resources to renewable energy has been attractive. This diversification is attractive due to the ethical attributes related to environmental qualities, and some renewable energy resources for the residential electricity consumption are cost-effective during periods of high oil prices (Krozer 2012a). The electricity growth in businesses was in line with the FOB prices. The gas consumption has grown very fast because policies promoted gas as the main substitute for coal and oil with regard to the mitigation of climate change and less dependence on the oil imports. The faster growth of the gross margins than the sales prices means higher energy consumption despite increasing energy prices. This indicates low price elasticity of demand for energy services. The price elasticity of demand for energy services is estimated to be between −0.6 and −0.9, which means price inelastic (though Eurostat data is imperfect for such estimates). The price inelastic demand implies that residents and businesses are not very sensitive to the supplied prices. This enables introduction of the costly innovations if these serve ethical attributes, such as environmental qualities due to renewable energy, social inclusion due to local sourcing of energy, energy security due to lower import dependence and so on. These can overcome the temporary high prices of innovations until the novel products and services attain price parity with the large-scale, fossil fuel-based ones. This does not hold for the electricity prices of businesses, which are estimated to be elastic. The emergence of the local energy initiatives for smart grid entailing energy service companies can largely be explained by combinations of the growing demands for renewable energy due to the ethical attributes and the high margins along with the price inelastic demand for the residential electricity consumption. The evolution of the local energy initiatives into energy service companies has robust economic rationality.

11.3 Support of Energy Consumption

Do policies foster or impede the 'smart grid'? Formally it is a policy priority but institutions can act stubbornly as mentioned in Chap. 7. It is often argued that the local energy initiatives are too small because economy of scale would be needed on energy markets; economy of scale means that the unit costs of an additional output increase because technologies are imperfectly divisible. If the economy of scale is valid, the large-scale production is lower cost per unit, and therefore, lower sales prices are sufficient to cover all costs compared to the small-scale production. Given the growing energy technology and demand, the large-scale suppliers would have an advantage because they would be able to sell at lower prices than the small-scale suppliers in the distributed energy systems. However, if the demands decrease or diversify, the large-scale suppliers bear higher costs of idle capacity than the small ones respectively more costs of tuning to the demanded diversity. In addition, the energy suppliers can innovate aiming for the better divisible technologies, which counteract the scale advantages. The mainstream economic theory, therefore, predicts that the scale effects are small on the efficient markets. The large price discounts due to the large-scale supplies would indicate inefficiencies caused by imperfect market competition (e.g. oligopolies) and be induced by policy interventions (e.g. high taxes). Herewith, the energy prices and taxes are analysed on the energy markets.

In the European Union, the consumption prices of electricity and gas vary across countries per year. In a year, three elements constitute the consumption price of an energy resource in a country. One element is the suppliers' prices, which vary per supplier, demand peaks during day and night and with respect to the energy resource prices. The second one is the price of the electricity and fuel networks, which is reasonably stable. The third element is the tax per energy unit on top, which fluctuates with respect to the annual budget needs. The share of these elements in the consumption prices varies per country and type of consumer; for the small-scale residential electricity consumers in the European Union, it is on the annual average, roughly one-third per element. The suppliers' prices and taxes are discounted for the large consumers. The discounting of the suppliers' prices and tax exemptions are related to the scale of energy consumption. The network prices can be neglected because they are more stable with respect to the scale. If the prices and taxes decrease with the scale, they encouraged additional energy consumption given the low price elasticity of demand for most energy products, and they impede entry of the innovating energy service companies whose services are initially smaller scale than the services of the vested suppliers and whose focus is on the distributed energy supplies.

The gas and electricity prices and taxes for the residential and business consumption in the European Union during the period 2008–2011 are calculated and related to the scale of energy consumption. The calculated prices and taxes are annual averages per scale categories and per country as defined in the energy statistics (Eurostat). The statistical data show several scale categories with prices and taxes per energy resource and per country, but there is no data on the total annual energy use per scale

Table 11.3 Prices and taxes with larger price discounts and tax exemptions for the larger gas and electricity uses

Price discount and tax exemption in €/kWh	Average price	Maximal price discount (%)	Average tax	Maximal tax exemption (%)
Gas, residential	0.068	−40	0.016	−28
Gas business	0.063	−44	0.012	−63
Electricity, residential	0.209	−29	0.053	−29
Electricity business	0.141	−23	0.015	−80

category. Table 11.3 summarises the results: the average price per energy unit with the maximum discount and the average tax per energy unit with the maximum tax exemption. Appendix Table 4 shows details per scale category. In all scales and energy resources in nearly all countries, the larger scale of energy consumption goes along with the larger price discounts and even larger tax exemptions. The cross-country correlations between the increases of price discounts and tax exemptions in relation to the consumption scale categories are high in nearly all scale categories (R^2 between 0.98 and 1).

The maximum price discounts are 23 % up to 44 % for the largest scale compared to the smallest scale. The price discounts for the large-scale consumption are covered by the higher prices of consumers without price discounts, which are mainly the residential and small enterprises. The high price discounts partly reflect the capital costs of the indivisible technologies under assumption of the maximum utilisation of production capacity, but there is idle production capacity on the energy markets. The price discounts on the efficient markets are smaller, for instance, the discounts on the food commodity markets are usually much lower than mentioned above. Those large price discounts solely due to the economy of scale are not plausible. Possibly the high tax exemptions induce the price discounts. The tax exemptions are larger (28 % up to 80 %) and they increase even more with the scale across nearly all scale categories, in nearly all countries during all years (the only exception is the largest scale of the residential gas consumers whose price discounts are larger than tax exemptions in a few but not all countries). The tax exemptions could be a result of the amalgamated lobbies of the large-scale business energy consumers who gain cheaper energy and the large-scale energy suppliers who earn the margins between the tax exemptions and price discounts. Given the country budget income, these tax exemptions add to the taxes imposed on the small-scale consumers, in particular the residential and small-scale companies: the small energy consumers pay for the large ones. This policy increases energy inefficiencies in economies despite the policy priority for 'smart grid' and it creates barriers of entry for innovators. Appendix Table 5 shows details per country.

The total annual tax exemptions are estimated based on the assumption about the energy consumption per scale category because this data is not available in the statistics. It is the price multiplied by the scale of consumption on the assumption that the energy consumption per scale category reflects the Poisson distribution. The estimates are shown in Appendix Table 6. Table 11.4 summarises the annual costs

Table 11.4 Markets of energy in the European Union

Annual average in the € million	Costs	Tax exemptions	Percent (%)
Gas, residential	90,021	19,072	21
Gas business	90,021	37,565	42
Electricity, residential	142,514	31,632	22
Electricity business	318,618	29,335	9
Total	641,175	117,603	18

of gas and electricity consumption, the annual tax exemptions and its share in the total costs. The average annual tax exemptions are nearly € 118 (USD 153) billion. This is 18 % of the annual average € 641 (USD 833) billion costs of all gas and electricity consumption on-site in the European Union. Gas has the highest tax exemption percent-wise. This implies that policies promote consumption of energy resources instead of highly valued energy products. The highest tax exemptions percent-wise are in Denmark, France and the Netherlands. The gas-producing countries provide the largest tax exemptions, such as the Netherlands. Nearly all price discounts and tax exemptions are generated by the fossil fuel supplies as gas is a fossil fuel and about 80 % of all electricity production is fossil fuel based.

11.4 Support of Energy Production

In addition to the larger tax exemptions for the larger scale of energy consumption on-site, there is a policy financial support of the energy production on-site and energy consumption in transport. It is assessed whether this financial support is mainly for fossil fuel or for renewable energy. This is done in the context of a widespread opinion as if the renewable energy production would need large subsidies to attain price parity with the energy production based on fossil fuels; the slogan 'windmills rotate on subsidies' is popularised. This argument persists despite studies that underpin a larger total policy financial support of fossil fuels and nuclear energy production than one of renewable energy in 2001 (EEA 2004), albeit lower per energy unit, and large financial support of the fossil fuel interests during decades in the last century which caused lack of efficiencies and innovativeness in the energy business because rent-seeking was more attractive (Storchmann 2005).

The financial support, herewith, is considered in a broad sense of all monetary transfers from the public to private resources. The best known transfers are from the public to private budgets which are generally called subsidies. These are grants, guarantees for loans, financing infrastructure, support of intermediaries and so on. These transfers are also called 'on-budget' because these can be found on the policy budgets. They are also called pre-tax because taxes include this support. Much financial support is not shown on the policy budget. This 'off-budget' support covers tax credits, exemptions and rebates for specific fuels, accelerated depreciation

Table 11.5 A compilation of studies on energy subsidies total and on-budget in the total

Billion USD	Fossil fuels		Renewable energy	
	2001	2010	2001	2010
World total[a]	N.A.	1,600[b]	N.A.	N.A
On-budget	199	380	9	N.A.
European Union total[c]	23.9 (24.6)[d]	N.A.	5.3	N.A.
On-budget	6.6	N.A.	0.6	N.A.
Netherlands total[e]	2.4	4.4	0	0.2
On-budget	0	0.3	0	0.1

[a]For 2001 de Moor (2001); for 2010 Clemens (2013)
[b]Excluding renewable energy and excluding subsidies for nuclear energy
[c]For 2001 EEA (2004) (Oosterhuis 2001)
[d]Including 1.0 billion euro for nuclear energy
[e]For 2001: Beers and van den Bergh (2009) for 2010: Visser 2011

allowances, exemptions from standards, resource rents, implicit infrastructure works, implicit income transfers, non-internalisation of externalities and so on. The off-budget support is also called post-tax because it does not involve tax payments (Valsecchi et al. 2009). The total support of fossil fuels is much larger than the on-budget support. For instance, the global on-budget support of fossil fuels, excluding the support of nuclear power, is estimated by the International Monetary Fund to be USD 480 (€ 380) billion, whereas the total policy support of energy is estimated to be about USD 1900 (€ 1461) billion (Clemens 2013). Table 11.5 shows results of a few studies for 2001 and 2010 on the global scale, in the European Union and the Netherlands. Although data is missing and data on the support of nuclear energy is rarely available, the studies indicate that the fossil fuel support has grown and is larger than the renewable energy support.

The European Union policy financial support of fossil fuels and renewable energy is analysed more in detail. The OECD statistical data is used to assess the support of fossil fuels (OECD 2014). This data is deficient. The data on the nuclear energy is not available and not all countries in the European Union are covered because only a few former Communist countries are OECD members. The data on the on-budget and off-budget support also varies. For example, the Netherlands would provide € 325 (USD 422) million support in 2010, but a study (de Visser et al. 2011) underpins for that year € 265 (USD 344) million on-budget and € 4345 (USD 5648) million off-budget. The off-budget support covers € 459 (USD 597) million tax exemptions for the domestic transport fuels and € 3886 (USD 5052) million for the international transport fuels. Even if the financial support of the international activities is excluded, the total support of the domestic activities is about 2.2 times higher than the ones disclosed in the OECD data. The country support of the domestic fossil fuel consumption, therefore, is corrected by the 2.2 factor, except for those countries that have disclosed such off-budget support in the OECD statistics. These countries are Denmark, Germany, France, Slovenia, Slovakia, Finland, Sweden and the United Kingdom. Note that all this financial support of fossil fuels is in addition

11.4 Support of Energy Production

Table 11.6 Subsidies for fossil fuels excluding nuclear energy and renewable energy in the European Union

mln €	2008	2010
Fossil fuel subsidies excluding nuclear energy ([a])		
Coal	5,436	5,010
Oil	14,460	15,516
Gas	6,306	5,324
Total	26,202	25,850
Total, including tax exemptions	34,953	36,717
Renewable energy based on feed-in tariffs		
Biomass	12,314	8,630
Hydro	1,542	3,497
Geothermal	–	1,080
Solar	2,564	8,614
Wind	8,613	15,024
Total costs of feed-in tariffs	25,033	36,845

([a]) No data about Bulgaria, Lithuania, Latvia, Cyprus, Malta and Romania

to the tax exemptions for energy supplies related to the scale on-site, which are estimated in the former paragraph.

The data on the policy financial support of renewable energy is also deficient. The main financial support is the feed-in regulations. These are price guarantees that oblige energy companies to accept the renewable energy deliveries to the grid and provide the possibility to demand a feed-in tariff for these deliveries. The price guarantee regulations and tariffs vary across the European countries and with respect to energy resources, per delivered volume and accepted volume, per year, and there are additional country-specific conditionalities for the feed-in payment. Only the minimum and maximum feed-in tariffs that can be demanded by the renewable energy producers are found on the EU - Energy Portal for 2008 and 2010 (EU 2014). Having only this data, the expenditure per country is estimated by multiplying the average of minimum and maximum tariff per renewable energy resource with the volume of that renewable energy resource for electricity generation in the European statistics. The feed-in policy has largely contributed to the growth of energy firms (cross-countries correlation between the numbers of firms and feed-in expenditures are high: $R^2 = 0.93$ for 2008 and $R^2 = 0.79$ for 2010).

Table 11.6 shows the results. The OECD data indicate that the on-budget support of fossil fuels excluding nuclear energy was about € 26.2 (USD 34.1) billion in 2008 and it was a similar support in 2010. When this figure is corrected for all support of the domestic businesses as mentioned above, the financial support was nearly € 35 billion (USD 45.5) in 2008 and € 37 (USD 48.1) billion in 2010 (it is not 2.2 of the on-budget support because several countries are excluded). By far the largest support is for oil consumption, which contradicts the policies aiming to reduce energy dependency on imports and mitigate climate change. The costs of the feed-in tariffs for renewable energy were about € 25 billion (USD 32.5) in 2008; a similar support has been estimated by the Council of European Energy Regulators (CEER 2013). This support was € 36.8 (USD 47.8) billion in 2010. By far the largest support in 2008 was purposed for biomass to farmers and in 2010 for wind. It does not mean that this support has been realised because of additional conditions in many countries. For example, in the Netherlands, the so-called SDE for renewable energy was € 1.3 (USD 1.7) billion on budget but only € 0.45 (USD 0.6) billion is spent, as mentioned in Chap. 10.

Taking these uncertainties into consideration, the total policy support of fossil fuels until 2010 was larger than the total support of renewable energy and was similar in 2010. The argument that the total financial support of renewable energy is larger than one of fossil fuels is incorrect.

11.5 Conclusions

The energy markets are assessed with regard to the mushrooming local energy initiatives. Many evolve into enterprises called energy service companies with the focus on the distributed energy systems called smart grid. These innovators can be considered signals of shifts on the energy markets from the bulky supplies to more customer-specific services. It is shown that during the period 2008–2011, which is after the economic crisis, the number of enterprises in the energy business increased by average 3600 with 23,000 jobs a year, which is more than in other businesses in the European Union. This growth is despite the stagnant total energy consumption in Europe. The main markets are substitution of fossil fuels for renewable energy and energy efficiency through value addition. Particularly the residential electricity consumption is attractive for innovating in the 'smart grid'. The energy service companies capture a growing market share because the vested energy businesses have not diversified into renewable energy and high-value services. This growth is despite policies in the European Union countries that impede these innovators. One major impediment is the larger tax exemptions for the larger-scale energy consumption. The tax exemptions cause that energy efficiency is unattractive and impedes energy services for the distributed systems that are prioritised in these policies. This policy protects the vested large-scale energy businesses. The tax exemptions in the European Union approximate € 118 billion a year, which is about 18 % of all energy costs. It has several consequences. In addition to the large tax losses, energy efficiency and mitigation of climate change are obstructed. This policy also encourages the rent-seeking behaviour instead of innovative responses to the changing demands because tax evasion is more economic than innovation. In addition, the policy financial support of the fossil fuels production was growing and it was larger than the support of renewable energy until 2010. That year, the fossil fuels were supported with € 37 (USD 48) billion a year and even more if the support of nuclear power is included, but this data is not found. The biggest part of the fossil fuel support is for oil, which is inconsistent with the European Union policy on energy independency because it stimulates oil imports and it obstructs the policy on climate change mitigation. About a similar amount is in support of all renewable energy though it is a larger support per energy unit. The sustainable innovators on energy markets can be enhanced when all policy support is abolished, in particular the tax exemptions for the energy production and consumption. This would create more transparent market and consistent energy policy.

Appendix

Table 1 Number of companies in electric power, gas and steam supply

	Total				Growth			
	2008	2009	2010	2011	2009 (%)	2010 (%)	2011(%)	Aver. (%)
EU 27	59,015	73,879	94,874	113,178	−75	−72	−81	24
Belgium	265	302	384	426	14	27	11	17
Bulgaria	528	1,166	1,412	1,802	121	21	28	57
Czech Republic	822	1,599	3,008	5,411	95	88	80	88
Denmark	1,646	1,660	1,674	1,792	1	1	7	3
Germany	23,445	28,765	38,821	48,284	23	35	24	27
Estonia	249	246	226	226	−1	−8	0	−3
Ireland	209	283	310	325	35	10	5	17
Greece								
Spain	14,346	14,990	15,319	15,593	4	2	2	3
France	4,493	10,034	16,403	17,473	123	63	7	64
Croatia								
Italy	2,478	2,973	4,097	6,601	20	38	61	40
Cyprus	2	6	9	0	200	50	−100	50
Latvia	281	324	373	405	15	15	9	13
Lithuania	252	292	405	518	16	39	28	27
Luxembourg	52	56	59	63	8	5	7	7
Hungary	489	506	556	585	3	10	5	6
Malta				3				
Netherlands	617	704	679	827	14	−4	22	11
Austria	1,791	1,861	1,922	1,976	4	3	3	3
Poland	2,332	2,826	3,200	3,596	21	13	12	16
Portugal	665	700	730	801	5	4	10	6
Romania	506	609	885	924	20	45	4	23
Slovenia	405	493	636	812	22	29	28	26
Slovakia	273	320	410	462	17	28	13	19
Finland	718	728	740	770	1	2	4	2
Sweden	1,546	1,681	1,791	2,083	9	7	16	11
United Kingdom	605	755	825	1,420	25	9	72	35

Table 2.a Final energy consumption in the European Union (EU) of 27 states, annual average during 1995–2011

1000 t.o.e. 11.63 GWh			On site		Transport	
	Total	t.o.e./person	Residential (%)	Business (%)	Passengers (%)	Freight (%)
EU total in t.o.e.	1,139,618	2.3	26	43	25	5
EU total in GWh	13,254	27.0	–	–	–	–
Belgium	36,698	3.5	25	48	21	6
Bulgaria	9,779	1.3	23	52	22	3
Czech Republic	25,495	2.5	25	55	16	4
Denmark	15,168	2.8	29	36	19	14
Germany	222,625	2.7	29	42	27	2
Estonia	2,780	2.0	35	40	20	5
Ireland	10,987	2.7	25	37	31	7
Greece	19,298	1.8	25	35	25	15
Spain	83,768	1.9	16	42	37	5
France	155,832	2.5	26	42	31	1
Croatia	5,775	1.3	30	39	29	1
Italy	124,999	2.1	23	43	26	8
Cyprus	1,727	2.3	13	34	49	4
Latvia	3,851	1.7	38	36	13	12
Lithuania	4,486	1.3	34	36	23	7
Luxembourg	3,844	8.3	13	30	54	3
Hungary	16,813	1.7	35	42	21	1
Malta	393	1.0	19	25	48	7
Netherlands	50,864	3.1	21	51	14	15
Austria	25,598	3.1	25	44	30	1
Poland	60,475	1.6	32	48	18	2
Portugal	17,439	1.7	17	45	32	6
Romania	24,749	1.2	32	50	14	3
Slovenia	4,677	2.3	24	43	31	2
Slovakia	10,825	2.0	22	59	18	1
Finland	24,981	4.8	21	61	13	5
Sweden	34,034	3.7	22	54	18	6
United Kingdom	147,432	2.4	29	36	29	7

Appendix

Table 2.b Energy consumption, residential mid-scale, average prices and trend during 2005–2011

	Electricity			Gas (kWh=0.036 GJ)		
	kWh/person	€/kWh	Price trend	kWh/person	€/kWh	Price trend
EU (27 countries)	1,632	0.10	106	71,741	0.001	116
Belgium	1,979	0.12	110	100,669	0.001	118
Bulgaria	1,339	0.05	98	1,664	0.001	116
Czech Republic	1,414	0.09	120	66,913	0.001	132
Denmark	1,867	0.10	112	39,172	0.002	102
Germany	1,716	0.12	98	87,540	0.001	112
Estonia	1,363	0.05	99	11,953	0.001	133
Ireland	1,807	0.13	115	44,919	0.001	128
Greece	1,591	0.07	117	6,016	N.A.	N.A.
Spain	1,518	0.10	116	25,603	0.001	110
France	2,332	0.08	98	68,101	0.001	118
Croatia	1,516	0.07	107	41,718	0.001	102
Italy	1,143	0.13	99	93,548	0.001	114
Cyprus	1,993	0.12	130	N.A.	N.A.	N.A.
Latvia	868	0.06	104	16,577	0.001	148
Lithuania	813	0.06	108	15,073	0.001	129
Luxembourg	1,665	0.12	103	130,199	0.001	126
Hungary	1,129	0.09	113	108,206	0.001	150
Malta	1,463	0.10	140	N.A.	N.A.	N.A.
Netherlands	1,472	0.12	109	144,169	0.001	112
Austria	2,090	0.11	114	47,518	0.001	119
Poland	707	0.08	105	27,541	0.001	122
Portugal	1,306	0.11	91	7,325	0.002	112
Romania	518	0.05	110	35,704	0.000	113
Slovenia	1,526	0.08	102	16,128	0.001	129
Slovakia	841	0.10	104	74,533	0.001	123
Finland	3,980	0.08	108	2,277	N.A.	N.A.
Sweden	4,296	0.10	116	2,301	0.002	117
United Kingdom	1,927	0.11	125	141,670	0.001	125

Table 2.c Energy consumption, business mid-size, average prices and trend during 2005–2011

	Electricity			Gas		
	kWh/GDP	€/kWh	Price trend	kWh/GDP	€/kWh	Price trend
EU (27 countries)	168,947	0.07	121	8,815,671	0.00	129
Belgium	180,079	0.08	126	415,780	0.001	135
Bulgaria	616,270	0.04	108	1,103,629	0.001	134
Czech Republic	339,972	0.08	146	510,461	0.001	137
Denmark	102,806	0.07	111	192,920	0.001	107
Germany	162,246	0.08	112	262,212	0.001	128
Estonia	362,004	0.04	100	597,824	0.001	168
Ireland	101,616	0.10	121	243,596	0.001	N.A.
Greece	168,749	0.07	116	153,476	N.A.	N.A.
Spain	179,879	0.08	128	309,014	0.001	144
France	151,301	0.05	110	166,253	0.001	130
Croatia	214,908	0.06	125	569,149	0.001	N.A.
Italy	156,197	0.09	117	387,993	0.001	125
Cyprus	180,666	0.11	153	–	N.A.	N.A.
Latvia	270,875	0.05	139	933,783	0.001	157
Lithuania	242,099	0.06	134	1,136,026	0.001	168
Luxembourg	166,109	0.08	117	326,906	0.001	132
Hungary	240,797	0.07	121	941,281	0.001	127
Malta	219,461	0.11	172	–	N.A.	N.A.
Netherlands	149,076	0.08	105	595,377	0.001	117
Austria	161,034	0.07	130	287,268	0.001	129
Poland	304,397	0.06	139	379,371	0.001	132
Portugal	208,836	0.07	113	256,836	0.001	127
Romania	325,840	0.05	93	1,353,125	0.000	126
Slovenia	295,319	0.07	126	283,905	0.001	160
Slovakia	404,894	0.08	146	976,114	0.001	154
Finland	362,980	0.06	109	257,524	0.001	113
Sweden	278,261	0.06	143	34,774	0.001	123
United Kingdom	118,953	0.08	151	349,643	0.001	121

Table 3.a Fuel prices and energy sales prices in EU countries 2004–2011

	€/kWh	€ mln	Growth		€/kWh	€ mln	Growth	
	Fuel to sale price €/kWh (%)	Gross margin	Sale price (%)	Gross margin (%)	Fuel to sale price €/kWh (%)	Gross margin	Sale price (%)	Gross margin (%)
	Gas consumption business				Gas consumption, residential			
Belgium	83	565	6	84	23	436	6	6
Bulgaria	127	(11)	9	−576	40	2	5	20
Czech Rep.	85	236	9	88	29	183	11	25
Denmark	91	124	10	86	20	113	9	22
Germany	60	5,567	9	14	21	3,271	3	0
Estonia	133	(6)	12	−99	47	(2)	10	−626
Ireland	132	324	−1	38	22	86	−2	6
Greece	0	(282)	0	29	0	(19)	0	83
Spain	92	929	8	157	22	489	1	4
France	75	1,680	9	37	23	1,773	5	4
Croatia	95	92	8	−58	39	21	3	143
Italy	79	2,812	5	36	24	2,114	4	2
Cyprus	0	–	0	0	0	–	0	0
Latvia	120	4	10	189	46	(0)	13	−503
Lithuania	96	67	11	−430	39	6	8	109
Luxembourg	68	76	8	16	26	22	8	21
Hungary	87	304	3	92	39	117	14	−189
Malta	0	–	0	0	0	–	0	0
Netherlands	78	1,598	3	20	23	978	5	1
Austria	75	420	6	93	23	163	5	8
Poland	86	393	9	−112	32	209	8	145
Portugal	79	212	7	29	19	43	3	12
Romania	166	(233)	5	−240	68	(67)	−1	−1494
Slovenia	79	41	14	54	25	12	8	14
Slovakia	85	168	8	41	32	82	7	33
Finland	78	216	4	224	0	(3)	0	22
Sweden	59	103	9	123	18	14	8	14
United Kingdom	84	2,516	5	−118	28	2,534	6	30

Table 3.b Fuel prices and energy sales prices in EU countries 2004–2011

	€/kWh	€ mln	Growth		€/kWh	€ mln	Growth	
	Fuel to sale price €/kWh (%)	Gross margin	Sale price (%)	Gross margin (%)	Fuel to sale price €/kWh (%)	Gross margin	Sale price (%)	Gross margin (%)
	Electricity consumption business				Electricity consumption, residential			
Belgium	24	2,098	2	3	6	1,227	3	−2
Bulgaria	47	249	2	1	15	169	0	2
Czech Rep.	24	1,416	10	12	8	565	7	7
Denmark	28	644	3	2	7	483	3	2
Germany	23	13,837	1	1	5	8,335	0	−1
Estonia	43	78	1	0	13	38	0	0
Ireland	20	735	4	4	5	494	5	7
Greece	28	999	3	3	10	532	5	6
Spain	25	5,700	8	8	7	3,139	6	10
France	35	6,058	3	2	8	5,645	0	−1
Croatia	31	228	6	9	10	196	3	3
Italy	21	9,412	3	3	5	4,394	−2	−2
Cyprus	18	135	10	13	6	87	8	12
Latvia	41	77	8	10	12	42	6	10
Lithuania	31	153	7	8	11	66	5	8
Luxembourg	24	193	3	2	6	49	0	1
Hungary	27	679	3	3	8	429	4	4
Malta	19	56	15	17	7	27	14	14
Netherlands	24	2,904	0	−1	6	1,313	2	1
Austria	26	1,347	6	7	6	867	4	4
Poland	31	2,238	10	13	8	987	5	6
Portugal	26	1,088	3	3	6	693	−4	−3
Romania	37	620	5	5	13	211	3	8
Slovenia	29	264	3	2	9	110	1	1
Slovakia	23	696	7	8	7	206	2	1
Finland	34	1,380	2	−1	8	807	2	2
Sweden	31	2,211	7	8	7	1,770	5	3
United Kingdom	25	7,031	9	8	7	5,862	5	3

Appendix

Table 4 Price discounts and tax exemptions in relation to scale of energy consumption

All in €/kWh	Prices excluding taxes versus scale				Taxes versus scale			
	2008	2009	2010	2011	Aver.	2008	2009	2010	2011	Aver.
Gas, residential										
<5,500 kWh	0.079	0.063	0.062	0.068	0.068	0.015	0.016	0.016	0.018	0.016
<55,500 (%)	−28	−31	−32	−31	−31	−19	−20	−21	−20	−20
<555,500 (%)	−9	−10	−10	−9	−9	−9	−7	−8	−7	−8
Total tax exemption (%)	−37	−41	−42	−41	−40	−28	−27	−28	−28	−28
Gas, business										
<0.288* 10^6	0.053	0.053	0.052	0.056	0.063	0.012	0.012	0.012	0.013	0.012
< 2.88* 10^6 (%)	−9	−12	−10	−7	−10	−13	−16	−15	−13	−14
< 28.88* 10^6 (%)	−11	−15	−15	−15	−14	−14	−17	−16	−16	−16
<288.88* 10^6 (%)	−12	−13	−14	−15	−14	−17	−17	−18	−19	−18
<1.111* 10^9 (%)	−6	−11	−8	−8	−6	−14	−18	−15	−14	−15
All tax exemption (%)	−38	−52	−48	−45	−44	−58	−68	−64	−63	−63
Electricity, residential										
<1,000	0.188	0.198	0.211	0.239	0.209	0.048	0.049	0.055	0.058	0.053
<2,500 (%)	−20	−20	−21	−24	−21	−23	−24	−24	−22	−23
<5,000 (%)	−5	−6	−6	−6	−6	−5	−5	−6	−6	−5
<15,000 (%)	−3	−3	−2	−3	−3	−1	2	−1	0	0
All tax exemption (%)	−27	−28	−29	−33	−29	−28	−28	−31	−28	−29
Electricity, business										
<20 MWh	0.120	0.140	0.150	0.154	0.141	0.013	0.012	0.017	0.017	0.015
<500 MWh (%)	−3	−10	−12	−12	−9	−55	−49	−48	−45	−49
<2,000 MWh (%)	−6	−6	−5	−7	−6	0	−17	−18	−17	−13
<20,000 MWh (%)	−8	−9	−6	−7	−8	−28	−16	−14	−11	−17
All tax exemption (%)	−7	−4	−5	−4	−23	−83	−82	−80	−73	−80

Table 5 Total annual tax exemptions and costs

In million euro	Tax exemption				Costs			
	Gas		Electricity		Gas		Electricity	
	Residential	Business	Residential	Business	Residential	Business	Residential	Business
EU	19,072	37,565	31,632	29,335	90,021	174,359	,514	318,618
Belgium	591	1,398	845	1,030	2,678	6,923	4,911	11,074
Bulgaria	4	163	102	17	24	1,132	872	1,996
Czech Rep	268	497	376	49	1,447	2,386	3,195	8,056
Denmark	408	1,870	1,084	1,986	614	2,637	2,977	4,977
Germany	5,963	7,174	10,673	11,572	26,024	40,267	40,761	77,829
Estonia	5	45	33	44	28	280	185	491
Ireland	62	243	514	19	389	1,698	713	3,269
Greece	–	–	254	368	55	689	1,622	5,070
Spain	422	1,874	2,533	1,220	1,932	7,540	18,211	34,190
France	1,878	2,470	4,910	–	12,776	13,318	–	–
Croatia	42	162	139	19	245	926	968	1,598
Italy	4,692	6,064	2,814	6,658	11,925	26,005	15,805	54,010
Cyprus	–	–	43	10	–	–	199	480
Latvia	7	104	15	–	49	330	210	654
Lithuania	16	191	34	15	89	1,126	289	887

Appendix

Luxembourg	13	33	16	52	120	433	175	826
Hungary	365	811	221	167	1,748	3,732	1,678	3,172
Malta	–	–	5	–	–	–	124	328
Netherlands	2,438	6,143	964	2,236	5,765	10,396	5,189	13,526
Austria	271	877	867	902	969	3,542	4,039	7,795
Poland	385	830	645	647	2,065	4,448	4,183	14,300
Portugal	14	136	773	1,020	190	2,117	3,374	5,969
Romania	324	1,237	159	–	963	3,578	1,164	3,911
Slovenia	20	104	112	67	72	354	581	1,487
Slovakia	177	364	105	16	1,115	2,238	876	3,912
Finland	–	–	738	208	7	886	4,241	6,345
Sweden	36	463	2,125	64	67	585	9,230	12,873
United Kingdom	672	4,312	531	947	18,667	36,794	16,740	39,594

Table 6 Assumed scale distribution for calculating total tax exemptions and costs

Gas				
Residents	<20 GJ	20–200 GJ	>200 GJ	
Share in total	0.6	0.3	0.1	
Business	<1000	1000–10,000	100,000–1,000,000	1,000,000–4,000,000
Share in total	0.4	0.3	0.1	0.05
Electricity				
Residential	<1000	1000–2500	5000–15,000	>15,000
Share in total	0.4	0.3	0.1	0.05
Business	<20 MWh	20–500 MWh	2000–20,000 MWh	>20,000 <70,000
Share in total	0.4	0.3	0.1	0.05

References

Arentsen, M., & Bellekom, S. (2014). Power to the people: Local energy initiatives as seedbeds of innovations? *Energy, Sustainability and Society, 4*(2), 2–12.
Bertoldi, B., Boza-Kiss, B., Panev, S., & Labanca, N. (2014). *ESCO market report 2013*. Ispra: Joint Research Centre, Mimeo.
Boon, F. (2012). *Local is beautiful*. Master thesis, University of Utrecht – Geoscience, Mimeo.
Clemens, B. (2013). *Energy subsidies reform: Lessons and implications*. International Monetary Fund, Mimeo.
de Moor, A. (2001). Towards a grand deal on subsidies and climate change. *Journal of Natural Resources Forum, JNRF, 25*(2), 167–176.
de Visser, E., Winkel, T., de Vos, R., Blom, M., & Afman, M. (2011). *Overheidsingrepen in energiemarkt*. Delft: CE en Ecofys.
Dragoman, M. C. (2014). *Factors influencing local renewable energy initiatives in different contexts*. Master thesis, University Twente, Mimeo.
EEA. (2004). *Energy subsidies in the European Union* (Technical Report 1/2004). Copenhagen, Mimeo.
EU. (2014). *Energy portal* http://www.energy.eu/country_overview/. Visited 18 Apr 2014.
Hielscher, S., Seyfang, G., & Smith, A. (2011). *Community innovation for sustainable energy*. Centre For Social and Economic Research on the Global Environment, School of Environmental Sciences, University of East Anglia Norwich, Mimeo.
Ison, N. (2010). *Governance of community energy projects*. Lancaster Environment Center, Mimeo.
Krozer, Y. (2012a). Costs and benefits of renewable energy in the European Union. *Renewable Energy, 50*(1), 68–73.
Krozer, Y. (2012b). Renewable energy in European regions. *International Journal on Innovation and Regional Development, 4*(1), 44–59.
OECD. (2014). http://www.oecd.org/site/tadffss/. Visited 18 Apr 2014.
Oosterhuis, F. (2001). *Energy subsidies in the European Union*. Amsterdam: Vrije Universiteit, Mimeo.
Ornetzeder, M., & Rohracher, H. (2006). User-led innovations and participation processes: Lessons from sustainable energy technologies. *Energy Policy, 34*, 138–150.

References

Rogers, J. C., Simmons, E. A., Convery, L., & Weatherall, A. (2008). Public perceptions of opportunities for community-based renewable energy projects. *Energy Policy, 36*, 4217–4226.

Sawin, J. L. (2013). Renewables 2013, Global Status Report, REN21, Paris.

Seyfang, G., & Haxeltine, A. (2012). Growing grassroots innovations: Exploring the role of community-based initiatives in governing sustainable energy transition. *Environment and Planning C: Government and Policy, 30*, 381–400.

Storchmann, K. (2005). The rise and fall of German hard coal subsidies. *Energy Policy, 33*(2005), 1469–1492.

van Beers, C., & van den Bergh, J. C. J. M. (2009). Environmental harm of hidden subsidies: Climate change and acidification. *AMBIO: A Journal of the Human Environment, 38*(6), 339–341.

Valsecchi, C., ten Brink, P., Bassi, S., Withana, S., M Lewis (IEEP), A Best, Rogers-Ganter, H., Kaphengst, T. (Ecologic), Oosterhuis, F. (IVM), & Dias Soares, C. (2009). *Environmentally harmful subsidies (EHS): Identification and assessment*. Brussels/London: Institute for European Environmental Policy.

Walker, G., & Devine-Wright, P. (2008). Community renewable energy: What should it mean? *Energy Policy, 36*, 497–500.

Chapter 12
Renewable Energy Business and Policy

What are the driving factors in renewable energy? Factors that impede or enhance the renewable energy business in the European Union are analysed. Nine factors are considered: population density, production output and energy sector output to indicate market conditions, public total expenditures, subsidies and environmental protection expenditures to indicate institutional conditions, R&D, share of students in population and venture capital to indicate firm's resources. Scarce space and vested energy interests are main barriers. R&D and venture capital are main drivers. The US and European Union support for R&D and venture capital in renewable energy are compared. The US support is larger and focused on the R&D grants. Large, innovative enterprises emerged. The European Union support is focused on the feed-in tariffs which reduce the investors' risks. Many small- and medium-size enterprises are generated. Regional policies can also reduce the investors' risks, which foster cost-effective renewable energy business. Policies can overcome barriers. The feed-in tariffs in the European Union proved to be cost-effective by the large renewable energy volume, many innovators and low public expenditures.

12.1 Renewable Energy Business

Given the demands for good environment shown in Chap. 3, it is discussed what factors drive renewable energy business and how policies can influence these factors. The starting point is that renewable energy businesses serve the private interests in energy consumption and the social interests for energy security, local sourcing, climate change mitigation and so on. The renewable energy business embraces firms in production, distribution and consumption of biomass and waste, hydropower, geothermal, solar and wind energy resources, as well as in energy efficiency through storage, distribution, cogeneration, processing and saving with related management and policymaking. The global investments in the renewable energy business grew during 2004–2010 on average 30 % a year to reach USD 273 (€ 211) billion in 2010 though the growth rates fluctuated from 0.4 to 75 % a year (McCrone 2012). This growth rate is higher than in most high-tech businesses. High growth is also expected in the next decades. Global scenarios generally assume a few percent higher energy growth rate than the income growth due to more energy demands in the emerging and low-income economies, and renewable energy is pursued to mitigate climate change and diversify resource for energy security (Statham 2007; Resch et al. 2008; Sawin and Moomaw 2009; Slot and Berg 2012). The subsequent energy scenarios done by international institutional and business organisations expect a higher share of renewable energy than the present 19 % of the present global energy consumption (Virdis 2003; van der Veer 2008). The scenarios made in commission of the non-governmental organisations envisage possibility of the wholly renewable energy consumption in the European Union by 2030 (Teske and Ties 2008) and globally by 2050 (Blok 2010). The international energy organisations, meanwhile, acknowledge that the renewable energy business becomes a dominant factor in the energy supply by 2040 (IEA 2015). Assuming that forecasting is not an 'objective science' but reflection of the institutional expectations and interests, the subsequent scenarios indicate that the renewable energy businesses can satisfy a large part of the global energy demands and growing global interests in this development.

Fostering the renewable energy business is the subject of studies from various angles. One is the managerial perspective, which is focused on the firm's internal and external resources for the socially beneficial renewable energy development as presented in Sharma and Vredenburg (1998). Another one is the neoclassic perspective that underpins the pivotal roles of prices for signalling deficiencies and allocating scarce resources, which is reviewed among others in Ruttan (2012). The institutional perspective addresses the decision-making processes with regard to technological, social, ethical and economic issues, which is reviewed in Steger et al. (2005). The evolutionary view is presented by Jacobsson and Johnson (2000) as a comprehensive framework with factors that are supposed to pose main barriers for renewable energy development. These factors are imperfect actors and markets (poorly articulated demand, established technology with increasing returns, local search processes, market control by incumbents), deficient networks (poor connectivity and wrong guidance with respect to future markets) and failing public institutions (legislative failures, failures in the educational system, skewed capital markets and underdeveloped organisational and political power of new entrants). Other authors argue that several drivers are essential: entrepreneurial activities, knowledge development, knowledge diffusion through networks, search guidance, market formation, resource mobilisation and creation of legitimacy (Hekkert et al. 2007). Several other studies are focused on balances of barriers and drivers and provide frameworks for this purpose (Foxon et al. 2005; Negro et al. 2007; Lund 2009). Herewith, this evolutionary framework is used to assess the main barriers and drivers for renewable energy with the statistical analysis and the institutional perspective is used to assess effects of the policies in the United States and European Union on the renewable energy business. These add to the neoclassic and behavioural perspectives on renewable energy in Chap. 11.

12.2 Barriers and Drivers

The factors that impede or enhance the renewable energy business are analysed statistically. Given the financial crash in 2008 with a dip in renewable energy investments, the analysis covers the period of continuous economic growth in the European Union and United States, which is during 1998–2008. Renewable energy is attractive during high fossil fuel prices. With regard to these prices, two periods are considered: 1998–2002 when the annual average real oil prices fluctuated at the level of the 2002 price and 2003–2008 when the real oil prices increased two times (the fuel prices are in USD2000, inflated with consumer price index and converted per year into euros). Nine factors are assessed with reference to the evolutionary framework of the barriers for renewable energy mentioned above and given the availability of the statistical data on the European member states excluding Malta and Croatia having poor data (Eurostat). These factors are population density, production output and energy output as proxies for the market conditions, public expenditures, subsidies and environment protection expenditures that refer to the institutional conditions, research and development (R&D) expenditures, students' share in population and venture capital that address firms' resources. Every factor is defined numerically and estimated for each country per capita per year. The annual results are compiled into the annual averages during 1998–2002 (low oil prices) and 2003–2008 (high oil prices). The missing data is extrapolated linearly. The largest renewable energy producers are not always the largest consumers. Hence, all producers and consumers and the ten largest ones are assessed. The aim was to assess the regional barriers and drivers as well. However, few regional data is found: only eight regions in Austria, which produced renewable energy above the EU average, and seven regions in Hungary, which produced little renewable energy. These are left out because they add little to the countries' assessment. The cross countries' data on each factor mentioned above are correlated with the renewable energy production and consumption using Pearson coefficients (R^2). Sensitivity analyses cover correlations per year and compiled into the average correlations and per energy resource: biomass and waste, hydro, geothermal, solar and wind. It is assumed that the regressions R^2 larger than 0.5 or smaller than −0.5 indicate important factors but not causal relations.

The results are summarised in Table 12.1. The columns show the renewable energy producers and consumers in the European Union and the ten largest ones. All are divided into periods 1998–2002 and 2003–2008. Appendix Tables 1, 1.a, 1.b, and 1.c show data per country: the annual average per person in MW renewable

Table 12.1 Factors that influence renewable energy in the European Union countries

Correlation are annual averages of the periods	All European Union		Ten largest		All European Union		Ten largest	
	1998–2002	2003–2008	1998–2002	2003–2008	1998–2008	2003–2008	1998–2008	2003–2008
	Production				Consumption			
MWh/person	2.3	3.0	7.4	8.3	1.0	1.1	7.0	7.8
Population density	(0.5)	(0.5)	**(0.6)**	**(0.7)**	(0.4)	(0.4)	(0.4)	(0.4)
Production volume	0.2	0.1	0.5	0.5	0.3	0.3	**0.6**	**0.6**
Energy output	0.3	0.0	(0.4)	**(0.6)**	0.3	(0.2)	0.0	(0.3)
Public expenditure	0.3	0.2	0.5	0.4	0.5	0.4	**0.6**	**0.6**
Subsidies' volume	0.5	0.3	0.3	0.3	0.5	0.5	0.5	0.5
Environment protection	0.3	(0.0)	0.1	0.2	(0.0)	(0.0)	0.5	0.5
R&D expenditures	**0.6**	0.5	**0.7**	**0.7**	**0.6**	**0.6**	**0.8**	**0.8**
Students' share	0.1	0.4	0.5	0.3	0.2	0.1	0.2	(0.1)
Venture capital	0.3	**0.7**	**0.8**	**0.7**	**0.7**	**0.8**	**0.9**	**0.8**

energy production and consumption and the factors. The first row shows the renewable energy production and consumption per person. There are large differences between countries: the ten largest producers per person are nearly three times larger than the European Union average and the ten largest consumers seven times larger than the average. The renewable energy production and consumption have grown much faster during the increasing oil prices than during the low oil prices: production growth was 6 % to 2 %, respectively, and consumption growth was 4 % and 1 %. Other rows show the impeding and driving factors. High positive or negative correlations of the factors and renewable energy are shown bold. The main impediments and drivers are discussed and so interactions between the factors.

Population density, defined as the number of people per square kilometre, indicates space scarcity for renewable energy. Given that the renewable energy resources have lower energy density than fossil fuel resources, more space would be needed for the state-of-the-art renewable energy technologies to meet all energy demands in densely populated countries, such as the United Kingdom (MacKay 2009) and the Netherlands (van Baal 2011). This factor analysis confirms that the limited space is an important impediment for the production, though it is not for the consumption. The sensitivity assessment with correlation per year confirms this finding. In particular, the biomass production is constrained by scarce space.

Production volume is indicated by the gross domestic product in euros per person. It is observed that the environmental technologies and renewable energy production were larger and grew faster in the high-income countries than in the low-income economies (Lanjouw and Mody 1996; Popp 1998). The factor analysis confirms these studies only insofar that the largest per person economies are also the largest renewable energy consumers. Sensitivity analysis underpinned that the countries' total production volume is moderately important for renewable energy production and consumption, though it is relevant for solar energy, for instance, in Austria and Germany.

Energy output is indicated by the index energy output in euro per person. One could expect that the large energy business enables to produce more renewable energy due to scale advantages. Gross et al. (2003), for instance, observed that the scale of renewable energy production in the 1990s was associated with decreasing unit costs albeit it is disputed if the scale of the renewable energy production is important compared to other factors (Farrell 2014). Hence, a positive correlation could be expected. However, the energy output is negatively correlated with the renewable energy production, though less negatively with its consumption. The sensitivity analyses per year and per renewable resource underpinned this finding. Low renewable energy production is found in the large energy-producing countries compared to their GDP, such as Belgium, Cyprus, the Netherlands, Poland and the UK. Possibly, the large energy producers do not care much about renewable energy because fossil fuel resources are available and the vested energy interests in the energy production may have obstructed renewable energy business.

The role of public expenditure is indicated by all government expenditure in euros per person. It is often argued that high government expenditures for the renewable energy production and consumption are necessary because these are in the development phase (Jacobsson and Bergek 2004). Given priority for the renewable energy in many countries, one would expect much renewable energy production and consumption in the countries with high public expenditures. An indication of it is the observation that renewable energy in the European Union grows twice as fast as in other OECD countries as a result of government expenditures (Blok 2006). High positive correlations would be expected but only moderate correlations are found for the renewable energy production though higher for the consumption. Big public spenders per person are often small renewable energy producers, for instance, Belgium, Ireland, Luxembourg, the Netherlands and the United Kingdom. The sensitivity analysis per year confirms this finding. The analysis per resource shows that the government expenditures are important for the solar energy consumption but hardly for the other renewable energy resources.

Subsidies are indicated by all subsidies in euros per person. All subsidies are included, which means the subsidies are in favour of the renewable energy business and ones in support of the fossil fuel businesses. The subsidies in support of fossil fuels were until 2008 larger than for renewable energy as shown in Chap. 11. Regarding the ambivalent allocation of subsidies, one could expect moderate correlations. This analysis confirms this expectation. Nevertheless, the subsidies for hydro and wind production are important.

Environmental protection is indicated by the expenditures on environmental protection in euros per person. High expenditures suggest political interest in renewable energy as a tool of environmental policy next to other instruments, such as emission trading (Menanteau et al. 2003). High correlations could be expected. However, all correlations are low, even somewhat negative for the consumption.

The sensitivity analysis and the resource-specific assessments confirm this finding. Environmental policies rarely foster renewable energy but can cause a trade-off between renewable energy and environmental performance.

Research and Development (R&D) is indicated by the total research and development expenditures per person. High correlation between R&D and total industrial production is found in the European Union (Smith 2000). Hence, high R&D could foster renewable energy and fossil fuel. Nevertheless, high positive correlations are found between R&D and the renewable energy production. It is even higher for the consumption. This is confirmed in the sensitivity analysis and for the consumption of hydro and solar energy. The renewable energy business is apparently highly knowledge intensive.

Students' share is the share of students in population. The argument is that higher education increases managerial awareness about sustainability (Wagner 2009; Meek et al. 2010). Hence, high share of students in populations could foster renewable energy. High correlations could be expected. However, these are low for the production and negligible, even somewhat negative, for the consumption. The sensitivity analysis and correlations per resource confirm the results. The high concentration of students is a negligible factor for the renewable energy business.

Venture capital is the available venture capital in euros per person. It indicates equity in firms. Much venture capital could foster renewable energy and fossil fuels. High positive correlation between venture capital and renewable energy production and consumption can be expected. The factor analysis confirms this. The sensitivity analysis also underpins this result. In particular, the biomass and hydro production and consumption benefit from venture capital.

The main impediments for the renewable energy business are scarce space and large energy business. The former is important because renewable energy has lower energy density than fossil fuels and equivalent energy consumption needs more space. The latter is relevant because the vested energy business is largely fossil fuel based and, therefore, rival to the renewable energy business. The policies cannot do much about the space use but foster international specialisation with renewable energy production in the thinly populated countries, e.g. the solar energy production for Europe on Sahara. The impediments caused by the large energy business can be reduced through changes in taxation as discussed in Chap. 11 but this is laborious. The main drivers are large R&D and venture capital. The renewable energy business is know-how intensive. Links between these factors are also assessed. Space scarcity is unlinked, e.g. the sparsely populated countries do not necessarily have low R&D or lack venture capital. Energy output is also unlinked though many energy-intensive countries are know-how extensive. Although the public expenditures and subsidies are moderately correlated with renewable energy, they are linked to R&D, which suggests that the high public expenditures and subsidies could foster renewable energy through more support of R&D. The R&D and venture capital are linked as mentioned in Chap. 10. The countries in the top ten R&D and ones in the top ten venture capital per person produce nearly three times more renewable energy than the European Union average. However, the

12.3 Policies and Renewable Energy Business

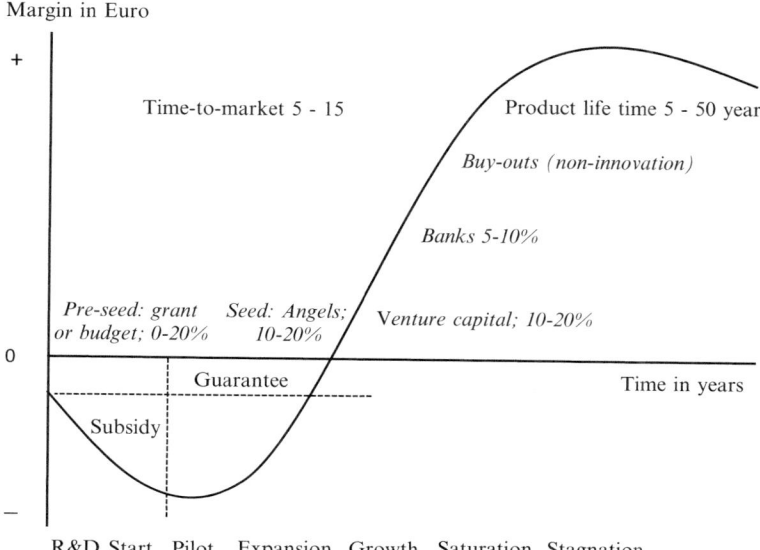

Fig. 12.1 Options for financing an innovation process

knowledge-intensive countries are not necessarily capital intensive as only 6 out of 12 countries are in the top ten countries for both factors. Policies aiming to strengthen these drivers are addressed.

12.3 Policies and Renewable Energy Business

The US and European Union public financial support of R&D and venture capital for the renewable energy businesses is discussed. These were the largest investors in renewable energy during the 2000s. The policy support is shown using the demand-pull model of innovations. Figure 12.1 illustrates it graphically. The figure presents the profit function of time. Time is on the horizontal axis and profit is on the vertical axis. Typical phases in this model, investors and usual interest rates are labelled on the figure. Dotted lines show options for the policies.

When an entrepreneur aims to launch a new product, it faces several years of costly research and development to demonstrate a saleable novelty (invention), followed by the start-up of an enterprise with pilot production and expansion of sales (innovation). Profit is negative during all these phases because costs are made without income. The innovators' costs increase during this so-called valley of death. This must be covered by the investors' equity, which is mainly based on public subsidies or private venture capital. The policymaking that aims to foster innovations has, in essence, only two options. Policies can reduce the costs through subsidies, in particular the most risky R&D phase. This option is shown in the figure by the area left of the vertical dotted line. The main instrument is subsidy. The innovators who get the subsidy benefit because their costs of innovation are reduced and they can attract venture capital because the costs of most risky phase are largely covered. This policy is specific because it supports the specific innovators and can be labelled as the supply-driven one because it induces supplies. The alternative policy is risk reduction due to the growing market. This is shown by the area beneath the horizontal dotted line in the figure. The main instrument is a price and volume guarantee for delivery on markets, which reduces the investors' risks because it enlarges markets without specifying whose risks are reduced due to the larger market. It is a generic support because it does not support any specific innovators and can be labelled as the demand-driven policy because fosters the innovations' adoption.

Various policy instruments based on these options are found. The United Nations Environment Programme (UNEP) presented instruments aiming to support R&D and to reduce the investors' risks (Makinson 2005), but many disappeared after the financial crisis in 2008. Another international review found 178 policy instruments, such as co-funding, soft loans, participation, grants, seed money, contracts, shares, credits and bonds. These are mainly grants for the R&D and soft loans for the market introduction. Per country, most instruments are found in the United States and Western Europe; only a few in the Eastern Europe, Asia, Australia/Oceania and South America and hardly any in Africa. Interviews with 31 innovators from different countries reveal that they use 163 instruments, which is average 5 per innovator but this number is higher in Western Europe and even higher in the United States (Ermen 2007). Detailed country analyses would reveal more instruments per country, e.g. 23 instruments are mentioned in the Netherlands alone in Chap. 10. These are instruments in the supply-driven policy. There are many instruments in the demand-driven policies, such as public procurement, education, labels, certifications, information, standards and norms and diffusion initiatives (after Johnson and Jacobsson 2002; Cunningham 2009; Iszak and Edler 2011), as well as the regional cluster and network policies, the emission trading and voluntary certification (COWI 2009). Also deposit refund, price guarantees and volume guarantees as well as take-back regulations can be added to because they enhance recycling. Liabilities can also be considered if consumers can demand for compensations of deficient products (Edler and Georghiu 2007; Vollebergh and van der Werff 2013). The main demand-driven instrument for the renewable energy business is the feed-in, a price guarantee for delivery of renewable energy to grid, which is introduced in Germany in 2004 (Böhme and Dürrschmidt 2007) and disseminated across the European Union with many adaptations with respect to resources, years of validity, delivery conditions and the tariff structure (Klein et al. 2008); the scale is covered in Chap. 11.

The effects of these policies on the renewable energy business are debated. Several studies argued that the high private investment in renewable energy with

12.3 Policies and Renewable Energy Business

large public support of R&D in the United States created its global superiority in the renewable energy business. This superiority is expressed as the largest and most innovative companies (Asplund 2008; Pernick and Wilder 2008; Siegel 2008; Rubino 2009). The US firms were the largest measured by stock market value in 2008 and covered about 42 % of the total USD 560 (€ 418) billion global value of the renewable energy business. They are considered innovative, though the European firms lead in wind, the Chinese in solar and the Japanese in energy-saving building and equipment. The competition assessments are compiled in Appendix Table 2. A few studies support this viewpoint, for instance, a Dutch study on the financing of renewable energy in the Netherlands, European Union and United States and it recommends more policy-specific support of demonstration projects (Biermans et al. 2009). A UK-based study also argued that the United States has generated more venture capital for the renewable energy business compared to the larger total venture capital in the European Union but the latter is mainly invested in the risk-avoiding acquisitions (Knight 2010). A study in the United States, however, argues that the policy support in the United States is too risk avoiding and more support of market introduction is advocated, which is the least risky phase of an innovation process (Zindler and Locklin 2010). Many advocate policy support of R&D on renewable energy and suggest that it is effective, particularly in combination with venture capital.

The feed-in regulations are also debated. The magazine *Economist* argued that the feed-in regulations invoke large market but distorts competition and that the US renewable energy business would be more innovative than the European Union one despite larger market because entrepreneurial capabilities, social mobility and venture capital are more important (Economist 2010). Many states in the United States, however, have introduced similar instruments. These are obligatory renewable energy purchases but without the feed-in tariffs, called the renewable portfolio standards. A few states and energy networks have also introduced the feed-in tariffs (Couture and Cory 2009). A US study on the European feed-in tariffs have not found competition-distorting effects and recommended similar regulation in the United States (Klein 2012). One can also be argued that the US and European Union policies are consistent with intentions. The narrative about the US intention is competitive business, which would need the focused, risk-taking R&D support with the aim to reach global innovation successes. The narrative about the European Union intention is consensus and risk-avoidance, which would need generic, instruments that enlarge the renewable energy markets. This way, both policies can be justified (as Rabbi Peter Tarlow says, 'to be sure of hitting the target, shoot first and call whatever you hit the target'). However, these narratives could also reflect successful lobbies of the vested scientific and business interests in the United States versus dispersed initiatives and firms across countries in the European Union.

It is also unclear whether the large and innovative firms in the United States emerged due to the R&D support when compared to the feed-in in the European Union or because the policy support of renewable energy after the financial crisis in 2008 was much larger in the United States than in Europe. The data suggest this.

Table 12.2 Number of enterprises, employees and their growth in the European Union and the United States; all are annual average, 2008–2011

		Number	Growth (%)
United States	Number of enterprises	12,634	1.9
	Employees	599,114	0.1
	Employees/enterprise	47	−1.8
European Union	Number of enterprises	85,237	24
	Employees	1,281,465	1.8
	Employees/enterprise	16	−18

The data is available about the policy support programmes aiming to overcome the financial crisis. The R&D support programme envisaged USD 94.1 (€ 70.2) billion in the United States compared to USD 22.8 (€ 17 billion) in the European Union (Biermans et al. 2009, p. 60). In addition, the European Union feed-in expenditures in 2008 were maximum USD 32.5 (€ 25) billion as shown in Chap. 11. The projected policy support of the renewable energy business was € 70.2 billion in the United States, nearly all supply-driven R&D support, compared to € 42 billion in the European Union, half of it demand driven. Another advantage of the renewable energy business in the United States was that the policy support of the rival fossil fuel business was lower because € 9.1 (USD 11.8) billion annual average compared to the minimum of € 25.6 (USD 33.3) billion in the European Union, both during 2005–2008 (OECD 2014); the latter is minimum because several countries and subsidies are not included. The larger public support of renewable energy in the United States than in the European Union provides an explanation for its large-scale innovative business and larger private investments in renewable energy. The renewable energy firms in the United States received more support than in Europe.

The issue is ultimately whether the US supply-driven policy focused on the specific R&D is more cost-effective compared to the European Union demand-driven policy with generic feed-in regulations after the crisis of 2008. The impact of the US support versus the European Union one is assessed with respect to the growth of renewable energy business. Table 12.2 shows the annual average number of enterprises and employees and their growth in the energy business in the United States and European Union during 2008–2011. The number of renewable energy business in the United States and its employment has hardly grown despite massive policy support, whereas this growth has been spectacular in the European Union.

Although not all new energy companies were focused on renewable energy, in particular in the United States many enterprises started operations in the shale gas, the differences in results are sufficiently large to underpin the impacts of policies. The large R&D support along with the risk-taking venture capital invoked fewer energy enterprises in the United States than in the European Union and their number hardly grows, but they are on average three times larger measured by the number of employees than the firms in the European Union. The feed-in tariffs invoked fast growth of the energy companies and generated twice as much employment mainly in small- and medium-size firms. Higher social benefits and lower social costs are attained in the European Union compared to the United States. The European Union policies are more cost-effective even though the US policies may have generated a few globally competitive firms.

The findings based on the statistical data are that the United States had much larger energy firms compared to the European Union before 2008 and its policy reinforced the scale. The European Union had smaller energy businesses and its policy has generated many more businesses and employment in small- and medium-size enterprises. The larger number of enterprises and jobs in the European Union is in contrast with its smaller policy support. Therefore, these socio-economic impacts are plausibly due to the different policy instruments. The successes of the demand-driven policy through the feed-in tariffs could have invoked pressures against this policy precisely because the innovating renewable energy business poses a serious competitive threat to the vested energy businesses.

12.4 Regional Policies

The energy policies are generally associated with the national regulations but many local and regional initiatives emerged as shown in Chap. 11. The regional policies can add to the development of the renewable energy business through the risk-reducing instruments. An illustration of a regional policy is presented not because this particular policy is successful as it shows failure of an inconsistent policy, but because the regional costs and benefits of renewable energy and policy instruments are underpinned given limited budgets. The context is the Energy Agreement signed in 2006 between the national and four regional authorities in an area branded as the Energy Valley in the Netherlands. This agreement envisaged 50 PJ of renewable energy and 5 million tons of CO_2 emission reduction in 2011 compared to 2006. The Frisian region within this area with roughly 30 % inhabitants would reach 13.5 PJ of renewable energy and 1.2 million tons of CO_2 emission reduction compared to the 2.8 PJ of renewable energy use and nil CO_2 emission reduction in 2006. The total fossil fuel consumption would be reduced by 21 % in this region. An investment programme is elaborated aiming to attain these regional objectives based on involvement of about 70 experts from businesses, authorities and institutes in several workshops. Possible actions in renewable energy and energy saving for CO_2 emission reduction (shown in italics) and investments in millions of euros (shown in parenthesis) are presented in Table 12.3. The actions would cover the household energy saving, transport with biogas and natural gas, wind energy on industrial parks, efficient greenhouses and biogas production from waste.

The present energy use, reduction of fuel use and CO_2 emission reduction, investment costs, capital costs, operational costs and cost savings due to the lower fuel use

Table 12.3 Possible Frisian actions for energy efficiency and renewable energy within the framework of the North Netherlands Energy Agreement

Energy consumption
Households: *insulation (447)*, heat pump plus storage (98), solar boilers (56), *micro-cogenerator (63)*, photovoltaic (157), *economy* light (17), CO_2 neutral dwelling (168)
Transport: *cars (50), biofuel and gas stations (15)*, hybrid cars (81), *gas for gasoline (244)*, SNG (244), CBG (244), biodiesel (49), *EU-CO_2 standard (98)*
Industries: wind energy on industrial parks on land (70), closed greenhouses (68), others (11)
Bio-residue processing to biofuels
Incineration for electricity and heat (150), technologies for biofuels (331), others (5)

are estimated for every action. Agreements between authorities and regional businesses underpinned the envisaged actions. The envisage investments are € 2.6 (USD 3.4) billion, which would require about € 650 (USD 845) investment per regional inhabitant per year. The expected annual costs and revenues are estimated at 5 % interest without subsidies based on the energy prices of 2006. This low interest rate is assumed possible due to the low risk investments based on a mix of public and private funding with guaranteed sales prices due to the feed-in policy. The annual costs would be € 0.3 billion but after subtracting cost savings due to lower energy consumption, net revenues of € 0.03 billion a year could be expected. About 27,000 jobs would be created during the 5 years of the programme, the regional businesses could gain about € 0.8 billion sales, and the region would benefit from better environmental qualities. The bottleneck was the financing because the national policy introduced subsidies per project instead of the feed-in policy. The risk-taking private investment would demand about 15 % return on investment entailing € 0.48 billion costs a year, and after subtracting the cost saving due to lower energy bills, about € 0.18 billion subsidies would be needed. Such large subsidies were considered unfeasible. The investment programme failed entailing the failure of the Energy Agreement.

Possible regional instruments to reduce the investors' risks are also discussed, given regulations on public investments and procurement in the European Union. Several financing and regulatory instruments in the regional policies are possible. The financing instruments can cover (a) infrastructure that encourages energy saving (e.g. local grid and biogas networks), (b) support of private investment through utilities (e.g. sludge processing of wastewater treatment), (c) local taxation differentiation (e.g. lower local tax for high energy saving), (d) co-funding of business clusters (e.g. promoting product sales), (e) participation in projects through regional development corporations (e.g. matching investments), (f) cofinancing of innovative projects (e.g. renewable energy technology), (g) European Union and national fund generation for the regional know-how (e.g. creating expertise centres) and (h) generating fund generation for upgrading regional skills (e.g. support of local energy initiatives). The regulatory instruments address public services and regulators. This can cover (a) public procurement guided by strict criteria (e.g. energy-saving public building, lighting, equipment), (b) tender specifications for constructions and equipment (e.g. public transport, lease of vehicles, performance contests), (c) user fees (e.g. parking lots, regulating waste processing), (d) promotional activities (e.g. contests, actions, quality scans, education, marketing) and (e) facilitation of licensing of renewable energy activities (e.g. flexibility for innovative solutions). This list of regional instruments is reasonably comprehensive compared to other assessments (e.g. Wientjes 2012). Such regional instruments need policy coordination across domains, which is risky because it needs a span of control across domains.

12.5 Conclusions

The renewable energy business is relevant not only to meet the global energy demands but also to generate income and jobs and mitigate environmental impacts, such as climate change. The question about how policies can foster renewable energy business is discussed based on the factor analysis of the renewable energy production and consumption with the European Union statistical data during 1998–2008. Among nine factors analysed, the main impediments for the renewable energy business are limited space and strong fossil fuel interests and the main drivers are large R&D and available venture capital. Policy support of the R&D and fostering venture capital are instrumental. The comparison of the US and European Union policies in support of the renewable energy business indicates that the United States provided much more public funding for renewable energy business. It is supply driven, mainly purposed for the private R&D. This policy has generated large, innovative firms. The European Union countries' policies invested less in the renewable energy business and focused on the price guarantees, which enlarge market without specification of the firms. This demand-driven policy has generated many more new enterprises and jobs than the larger funding in the United States. The US policies on renewable energy business are more risk taking and specific than in the European Union but the social benefits of the European Union policy are higher and the social costs of this policy are lower. The regional policies can add to development of the renewable energy business if they reduce the risks of entering renewable energy markets. Many financing and regulatory instruments are available and can be implemented if the authorities coordinate across various policy domains. The renewable energy business can be strengthened cost-effectively due to the demand-driven policy such as the feed-in regulation and the regional financing and regulatory instruments.

Appendix

Table 1 Per person renewable energy production, consumption and growth

42 GJ = 11.6 MWh	Production renewable energy		Production growth (%)		Consumption electricity generation		Consumption growth (%)	
	1998–2002	2003–2008	1998–2002	2003–2008	1998–2002	2003–2008	1998–2002	2003–2008
EU (27)	2.34	2.98	1.8	6.4	1.0	1.1	1.3	4.4
Belgium	0.61	1.07	5.2	**16.2**	0.2	0.3	3.8	**10.7**
Bulgaria	1.04	1.56	11.3	4.1	0.4	0.6	3.8	**6.0**
Czech Rep.	1.62	2.35	5.4	**6.9**	0.3	0.5	6.4	**8.2**
Denmark	3.85	5.52	6.1	5.5	1.2	1.8	12.7	5.5
Germany	1.29	2.91	9.6	**17.5**	0.5	1.0	11.2	**11.7**
Estonia	4.51	6.01	−0.1	5.5	0.5	0.8	0.5	**7.6**
Ireland	0.72	1.06	6.6	**11.5**	0.4	0.7	10.0	**12.4**
Greece	1.46	1.72	−0.1	2.7	0.5	0.7	1.1	**7.6**
Spain	2.02	2.49	0.2	5.5	1.1	1.4	2.1	**9.0**
France	3.07	3.00	−2.5	2.9	1.5	1.3	−0.5	1.8
Croatia								
Italy	1.91	2.25	2.5	4.9	0.9	0.9	1.3	2.4
Cyprus	0.74	0.85	−0.4	**7.2**	0.0	0.0	1.0	**16.6**
Latvia	7.59	9.11	4.1	2.7	2.0	2.2	1.0	4.2
Lithuania	2.27	3.13	7.1	5.9	0.5	0.6	3.9	5.7
Luxembourg	1.01	1.68	3.0	**14.2**	2.3	2.3	1.7	−0.4
Hungary	0.96	1.39	0.6	**11.2**	0.1	0.2	14.4	**12.5**
Malta								
Netherlands	0.99	1.41	8.4	**6.8**	0.2	0.4	8.2	**12.2**
Austria	9.40	10.12	1.8	3.8	5.7	5.6	2.3	0.8
Poland	1.19	1.42	1.3	4.6	0.2	0.3	1.3	2.4
Portugal	4.18	4.52	−1.2	3.7	1.5	1.6	−2.2	**11.5**
Romania	2.10	2.54	1.1	**6.5**	0.9	1.0	−2.2	4.2
Slovenia	3.92	4.48	5.7	2.9	2.1	2.1	3.1	4.2
Slovakia	1.25	1.84	10.7	6.0	1.0	0.9	5.5	−0.7
Finland	16.86	18.93	3.9	2.5	4.3	4.6	0.5	**5.9**
Sweden	18.17	18.30	1.9	2.6	9.6	8.6	−0.4	1.8
United Kingdom	0.45	0.70	6.0	**9.8**	0.2	0.3	6.8	**10.3**

The largest ten producers and consumers are underlined and the fastest ten growers are bold

Appendix

Table 1.a Factors per person: market conditions

	Population density person per km²		Gross domestic product in euro		Energy production index in euro	
	1998–2002	2003–2008	1998–2002	2003–2008	1998–2002	2003–2008
EU (27)	112	115	18,837	23,142	95	96
Belgium	334	342	24,319	29,710	55	96
Bulgaria	73	70	1,710	3,337	54	110
Czech Rep.	130	130	6,381	11,120	89	102
Denmark	124	127	32,100	39,143	67	92
Germany	230	231	24,979	27,873	94	98
Estonia	32	31	4,534	9,330	88	110
Ireland	54	59	27,543	39,998	–	
Greece	91	92	,750	18,225	88	96
Spain	81	88	15,693	21,668	83	90
France	110	115	23,623	27,982	92	93
Croatia						
Italy	189	194	20,942	25,039	86	95
Cyprus	75	82	14,344	18,831	76	91
Latvia	37	36	3,403	6,822	85	110
Lithuania	54	53	3,496	6,902	71	98
Luxembourg	167	179	48,739	69,364	82	90
Hungary	110	108	5,280	8,910	87	104
Malta	1,216	1,278	10,467	12,470		62
Netherlands	425	437	26,145	32,568	55	95
Austria	95	98	25,697	30,825	84	102
Poland	123	122	4,774	6,929	99	104
Portugal	115	119	12,062	15,013	94	91
Romania	94	91	1,840	4,284	79	103
Slovenia	98	99	10,956	15,365	85	95
Slovakia	110	110	4,108	8,215	93	98
Finland	15	16	25,298	31,305		
Sweden	20	20	28,240	34,221	101	92
United Kingdom	241	247	26,043	30,649	113	96

Table 1.b Factors per person: institutional conditions

	Total expenditure in euro		Subsidies in euro		Environmental protection expenditure in euro	
	1998–2002	2003–2008	1998–2002	2003–2008	1998–2002	2003–2008
EU (27)	8,713	10,760	246	265	131	146
Belgium	12,549	15,291	295	492	143	156
Bulgaria	777	1,427	21	31	6	17
Czech Republic	3,184	5,085	168	202	35	57
Denmark	18,123	21,264	784	873	269	252
Germany	12,028	12,996	409	316	170	163
Estonia	1,847	3,684	50	80	6	33
Ireland	9,970	15,123	208	200	221	383
Greece	6,152	9,143	18	18	79	120
Spain	6,482	8,989	171	222	40	75
France	12,743	15,330	359	403	128	158
Croatia						
Italy	10,325	12,298	254	253	183	208
Cyprus	5,874	8,409	163	132	31	56
Latvia	1,380	2,889	32	56	3	25
Lithuania	1,415	2,701	31	51	4	43
Luxembourg	20,560	28,490	756	1,086	303	321
Hungary	2,897	4,651	92	124	17	47
Malta	4,684	5,667	189	259	41	182
Netherlands	12,530	15,565	380	411	431	498
Austria	13,708	15,983	850	1,051	180	232
Poland	2,169	3,228	23	40	33	26
Portugal	5,526	7,059	150	135	75	86
Romania	721	1,803	28	56	5	19
Slovenia	5,396	7,263	206	257	57	125
Slovakia	2,032	3,425	93	122	15	28
Finland	13,022	16,122	383	416	153	170
Sweden	16,416	18,105	483	489	73	121
United Kingdom	10,799	13,702	118	189	137	176

Table 1.c Factors: business resources per person

	R&D total expenditure in euro		Students in total population (%)		Venture capital in euro		
	1998–2002	2003–2008	1998–2002	2003–2008	1998–2002	2003–2008	
EU (27)	308	414	3.2	3.7 %	78	154	
Belgium	432	537	1.4	3.7 %	50	44	
Bulgaria	7	14	3.1	3.1 %			
Czech Rep	62	134	2.3	3.1 %	4	6	
Denmark	627	948	3.5	4.0 %	56	38	
Germany	545	685	2.5	2.7 %	34	41	
Estonia	27	78	3.5	4.9 %			
Ireland	275	470	4.0	4.5 %	346	34	
Greece	55	103	3.7	5.4 %	7	8	
Spain	118	232	4.5	4.3 %	29	22	
France	460	581	3.4	3.4 %	74	62	
Croatia							
Italy	191	273	3.2	3.4 %	64	37	
Cyprus	33	69		2.6 %			
Latvia	11	32	3.4	5.4 %			
Lithuania	16	45	3.1	5.3 %			
Luxembourg	839	1,080	0.5	0.0 %	1,026	117	
Hungary	33	80	2.7	4.1 %	145	162	
Malta		57		2.2 %			
Netherlands	442	591	3.0	3.4 %	7	7	
Austria	487	696	3.2	3.0 %	74	19	
Poland	26	35	3.6	5.4 %	4	3	
Portugal	71	126	3.6	3.7 %	19	2	
Romania	7	16	1.8	3.4 %	14	14	
Slovenia	131	205	3.8	5.4 %			
Slovakia	27	36	2.3	3.3 %			
Finland	705	1,037	5.0	5.7 %	305	282	
Sweden	939	1,216	3.5	4.6 %	1,498	1,105	
United Kingdom	412	532	3.4	3.8 %	13	4	

Table 2 The main renewable energy market innovators in the United States (USA), EU and other countries

Businesses	Resources (input)	Products (output)	Firms' stock market value by 27-6-2008, USD billion (Rubino 2009)			Innovators (firms to watch, Pernick and Wilder 2008)		
			US	EU	Other	US	EU	Other
Biofuels	Oils	Biodiesel	NA	NA	NA	7	3	0
	Sugars	Ethanol						
	Waste	Biogas						
Hydropower	Inland	Electric	–	–	–	–	–	–
	Waves and tide	Storage						
Geothermal	Groundwater	Heat pumps	2.2	0.5	1.1	–	–	–
	Deep							
Wind power[a]	Onshore	Electric	29[b]	153	9	5	4	1
	Offshore							
Solar power	Photovoltaics	Electric	31	27	80	6	1	3
	Thermal (CSP)	Heat						
Green buildings	Architecture	Storage	39[c]	1	49	7	0	3
	Lighting	Certification						
	Microgenerators							
Personal transport	Hybrid (electric)	Batteries	111	0.3	0	6	0	4
	Electric	Fuel cell						
	Hydrogen	Flywheels						
	Hybrid (air)	Compression						
Smart grid	Monitoring	Metres	24[d]	0.1	0	10		
	Point of use	Storage						
	Networks	Smart grid						
	Cogenerator	Heat reuse						
Appliances (mobile)	Photovoltaics	Embedded systems				7	2	1
Carbon trading	CO_2 emissions	Trading houses	0	2	0			
Total			237	184	139	48	10	12

Based on Asplund (2008), Pernick and Wilder (2008), Siegel (2008), and Rubino (2009)
[a]Excluding sails and kites for motion
[b]Assumed 10 % of the total general electric stock value USD 261 000 million
[c]Assumed 10 % of the Procter & Gamble (Duracell) USD 184 650 million
[d]Assumed 10 % of the IBM USD 164 900 million

References

Asplund, R. W. (2008). *Clean energy* (1st ed.). New York: Wiley.
Biermans, M., Grand le, H., Kerste, M., & Weda, J. (2009). *De Kapitaalmarkt voor Duurzame Projecten*. Amsterdam: SEO.
Blok, K. (2006). Renewable energy policies in the European Union. *Energy Policy, 34*, 251–255.
Blok, K. (2010). *The energy report*. Gland: World Wildlife Fund, Ecofys, OMA.
Böhme, D., & Dürrschmidt, W. (2007). *Renewable energy sources in figures – national and international development: Status 2007*. Federal Ministry for the Environment, Nature Conservation and Nuclear Safety, Berlin, Mimeo.
Couture T., & Cory, K. (2009). *State Clean Energy Policies Analysis (SCEPA) project: An analysis of renewable energy feed-in tariffs in the United States*, Technical report, NREL/TP-6A2-45551 Revised June 2009. Golden, Colorado.
COWI. (2009). *The Potential of Market Pull Instruments for Promoting Innovations with Environmental Characteristics*. European Commission, Directorate General Environment, Brussels.
Cunningham, P. (2009). *Demand-side innovation policies*. Pro Inno Europe, Mimeo.
Economist. (2010, June 10). *Europe's tech entrepreneurs*.
Edler, J., & Georghiu, L. (2007). Public procurement and innovation – Resurrecting the demand side. *Research Policy, 36*, 949–963.
Farrell, J. (2014). *Renewable energy economies of scale are bullshit, Institute for Local Self-Reliance*. http://www.ilsr.org/renewable-energy-economies-scale-are-bullshit/. Visit 13 Mar 2014.
Foxon, T. J., Gross, R., Chase, A., Howes, J., Arnall, A., & Anderson, D. (2005). UK innovation systems for new and renewable energy technologies: Drivers, barriers and systems failures. *Energy Policy, 33*(16), 2123–2137.
Gross, R., Loach, M., & Bouwen, A. (2003). Progress in renewable energy. *Environment International, 29*, 105–122.
Hekkert, M. P., Suurs, R. A. A., Negro, O., Kuhlman, S., & Smits, R. E. H. M. (2007). Functions of innovations system: A new approach for analysing technological change. *Technological Forecasting & Social Change, 74*, 413–432.
IEA, International Energy Agency. (2015). *World Energy Outlook 2014*, Paris.
Iszak, K., & Edler, J. (2011) *Trends and challenges in demand-side innovation policies*. Technopolis group, Mimeo.
Jacobsson, S., & Bergek, A. (2004). Transforming the energy sector: The evolution of technological systems in renewable energy technology. *Industrial and Corporate Change, 13*(5), 815–849.
Jacobsson, S., & Johnson, A. (2000). The diffusion of renewable energy technology: An analytical framework and key issues for research. *Energy Policy, 28*(9), 625–640.
Johnson, A., & Jacobsson, S. (2002). *The emergence of a growth industry: A comparative analysis of the German, Dutch and Swedish Wind Turbine Industries*. http://www.druid.dk/conferences/winter2002/gallery/jacobsson.pdf. Visited 30 Aug 2014.
Klein, A., Pfluger, B., Held, A., Ragwitz, M., Resch, G., & Faber, Th. (2008). *Evaluation of different feed-in tariff design options – Best practice paper for the International Feed-In Cooperation*. Fraunhofer Institut, Institut for Systems and Innovation Research, Mimeo.
Klein, C. A. (2012). *Renewable energy at what cost? Assessing the effect of feed-in tariff policies on consumer electricity prices in the European Union*. Thesis published in The Georgetown Public Policy Review, 18 (1), pp. 43–62.
Knight, E. R. W. (2010). *The economic geography of clean tech venture capital* (Working Paper). University of Oxford, Mimeo.

Lanjouw, J. O., & Mody, A. (1996). Innovation and the international diffusion of environmentally responsive technology. *Research Policy, 25*, 549–571.
Lund, P. D. (2009). Effects of energy policies on industry expansion in renewable energy. *Renewable Energy, 34*(1), 53–64.
MacKay, D. (2009). *Energy without hot air.* www.withouthotair.com
Makinson, S. (2005). *Public Finance Mechanism to Catalyze Sustainable Energy Sector Growth*, United Nations Environmental Programme (UNEP) and Sustainable Energy Finance Initiative (SEFI), Mimeo.
McCrone, A. (2012). *Global Trends in Renewable Energy Investments*, Bloomberg New Energy Finances. Frankfurt School, Mimeo.
Meek, W. R., Pacheco, D. F., & York, J. G. (2010). The impact of social norms on entrepreneurial action: Evidence from the environmental entrepreneurship context. *Journal of Business Venturing, 25*, 493–509.
Menanteau, P., Finon, D., & Lamy, M.-L. (2003). Prices versus quantities: Choosing policies for promoting the development of renewable energy. *Energy Policy, 3*, 799–812.
Negro, S., Hekkert, M., & Smits, R. E. (2007). Seven typical system failures that hamper the diffusion of sustainable energy technologies. *Energy Policy, 35*, 925–938.
OECD. (2014). http://www.oecd.org/site/tadffss/. Visited 18 Apr 2014.
Pernick, R., & Wilder, C. (2008). *The cleantech revolution* (1st ed.). New York: Harper.
Popp, D. (1998). *Induced innovation and energy prices.* Kansas University, Mimeo.
Resch, G., Held, A., Faber, T., Panzer, C., Toro, F., & Haas, R. (2008). Potentials and prospects for renewable energies at global scale. *Energy Policy, 36*, 4048–4056.
Rubino, J. (2009). *Clean money, picking the winners in the green-tech room* (1st ed.). New York: Wiley.
Ruttan, V. W. (2012). Source of technical change: Induced innovation, evolutionary theory, and path dependency. In A. Grübler, N. Nakicenovic, & W. D. Nordhaus (Eds.), *Technological change and the environment, resources for the future Washington DC* (pp. 9–39). Laxenburg: International Institut for Applied Systems.
Sawin, J. L., & Moomaw, W. R. (2009). *Renewable revolution: Low carbon energy by 2030.* Danvers: Worldwatch Institute.
Sharma, S., & Vredenburg, H. (1998). Proactive corporate environmental strategy and development of competitively valuable organizational capabilities. *Strategic Management Journal, 19*, 729–753.
Siegel, J. (2008). *Investing in renewable energy* (1st ed.). New York: Wiley.
Smith, K. (2000). *What is the 'knowledge economy'? Knowledge intensive industries and distributed knowledge bases.* Project "Innovation Policy in a Knowledge-Based Economy", European Commission, Mimeo.
Statham, B. A. (2007). *Deciding the future: Energy policy scenarios till 2050.* World Energy Council.
Steger, U., Achterberg, W., Blok, K., Bode, H., Frenz, W., Gather, C., Hanekamp, G., Imboden, D., Jahnke, M., Kost, M., Kurz, R., Nutzinger, H. G., & Ziesemer, T. (2005). *Sustainable development and innovation in the energy sector* (pp. 211–222). Berlin/Heidelberg: Springer.
Teske, S., & Ties, F. (2008). *Energy [R]evolution, greenpeace international Europe.* Amsterdam: Mimeo.
van Baal, M. (2011). *Kunnen wij overschakelen op duurzame energie.* Technisch Weekblad. http://www.technischweekblad.nl/rubrieken/energieserie/kunnen-we-overschakelen-op-duurzame-energie.166168.lynkx. Visited 15 Nov 2013.
van der Slot, A.,& van den Berg, W. (2012). *Clean economy, living planet.* Roland Berger Strategy Consultant.
van der Veer, J. (2008). *Shell energy scenario's to 2050.* Shell International BV, Mimeo.
van Ermen, R. (2007). *Comparison and assessment of funding schemes for development of new activities and investments in environmental technologies* (Fundetec, Report Number 044370), Brussels.

References

Virdis, M. R. (2003). *Energy to 2050*. Paris: IEA and OECD.

Vollebergh, H. J., & van der Werff, E. (2013). *The role of standards in eco-innovations: Lessons for policymakers*. Milano: FEEM 74.2013.

Wagner, M. (2009). Eco-entrepreneurship: An empirical perspective based on survey data. In G. D. Libecap (Ed.), *Frontiers in eco-entrepreneurship research* (Advances in the study of entrepreneurship, innovation & economic growth, Vol. 20, pp. 127–152). Bingley: Emerald Group Publishing Limited.

Wientjes, R. (2012). *Regional Innovation Monitor, demand side innovation policies at regional level*. UNU-MERIT, Mimeo.

Zindler, E., & Locklin, K. (2010). *Crossing the valley of death*. Bloomberg New Energy Finance.

Chapter 13
Conclusions on Sustainable Innovations

The perspective of meeting demands for income and good environment culminated in the ideal of income growth, fair distribution of wealth and availability of environmental qualities referred to as sustainable development. Sustainable innovations are discussed, considered novel technology uses that generate welfare growth, in particular income and better environmental qualities. A fair distribution of wealth is a precondition for these. Herewith, the findings from the policymaking perspective are reviewed on the assumption that the successful innovations are unpredictable and that neither the market nor policy is the panacea. Public institutions can execute routines very well but are often risk avoiding, and market forces act usually myopic. Less regulations and managerial interventions but far-reaching individual and social demands for environmental qualities with freedom of actions and prices that reflect social costs of environment could generate sustainable innovations. Above all, policies are about creating conditions for the diversity of innovating tinkerers, artists, inventors, engineers, researchers, managers and entrepreneurs that pursue sustainable innovations in firms, social organisations and institutions. Economic growth and environmental qualities are positively linked when the diversity of innovators is fostered and collide when sustainable innovations are suppressed by the rent-seeking behaviour.

13.1 Innovations for Sustainable Development

A conventional train of thought that environmental impacts increase with the population and income growth cannot stand scrutiny of the empirical observations. Along with the global real income growth, the material use per income has decreased throughout the last hundred years. Nearly half of the high-income countries whose income and private consumption value has grown during the last decades have reduced their total material use and environmental impacts, such as greenhouse gases, NOx emissions and water quality albeit waste and biodiversity issues aggravated. The decoupling of income from environmental impacts cannot be explained by the lower material use due to higher real natural resource prices because it is the decreasing trend, not by imports of products from the material-intensive production in the low-income countries because much bulky trade evolves between the high-income countries though these imports are relevant in some cases. It is also unrelated to environmental policies though these policies are relevant for the reduction of some impacts in some countries. The explanation is the technological change that evolves largely autonomously of policymaking. This change consists of three types of innovations: the labour for material substitution in production and consumption (services), the cost-saving material reduction in production (process innovations) and the value-augmenting labour addition to products and services (product innovation). The substitution has been observed in production as being the service growth, but the share of services in consumption increases slowly when measured by the household expenditures. In the high-income countries, the process innovations and product innovations are important for policies that pursue sustainable development. The service growth is relevant in the low-income countries. The decoupling is largely due to the economic circulations in which innovation rents of the cost-saving material reduction are allocated into the value-adding labour. The scale of the allocation approaches 13 % of the global GDP, which is sufficiently large to drive the decoupling. The innovation-based income growth reduces autonomously environmental impacts albeit slowly. Prime factor in this economic circulation is the know-how creation, as well as its use and costs. Policies that foster education, knowledge, creativity and skills and create conditions for interactions between various stakeholders generate know-how, which fosters sustainable innovations development.

Innovations can also be induced. Although innovations are rarely achieved on demand, the social sense of urgency about environmental qualities attracts activities that eventually generate sustainable innovations. The demands for environmental qualities increase. This is largely due to the growing knowledge work because the knowledge workers perform in good cultural and natural environments, because the growing leisure time needs good environment and more people can pay for leisure and because communication tools enable to disseminate information and opinions. These demands create markets for sustainable innovations. In addition to the conventional markets of material and energy efficiency and pollution controls, the emerging markets are the natural blends. The natural blends are attributes of environmental qualities, being collective goods expressed in the privately owned, man-made products and services. These expressions are tangible in the ethical consumption, interaction in nature and in cultural expressions. An indicative assessment shows that the global market of sustainable innovations is minimum USD 2,981 (€ 2,293) billion, which is about 4.6 % of the global GDP. For comparison, it is about half of all health-care expenditures. The growing demand for environmental qualities is the good side of the sustainable development coin. The downside is the policy financial support of the rent-seeking vested interests that harm environment and compete with the sustainable innovators. On the global scale, this support is even larger than the market of sustainable innovations due to demands for environmental qualities. Among the largest policy support are tax exemptions and subsidies for fossil fuels, polluting agriculture, payments for the wasteful infrastructure and degrading land use. A few of these impediments are assessed in monetary terms.

They add up to USD 3,053 (€ 2,349) billion a year. In addition, there are nonmonetary entitlements for the rent-seeking behaviour, e.g. permits, patents, copyrights and so on. The sustainable innovators are obstructed by the policy support of the environmentally harmful rivals. Hence, sustainable development moves slowly. Removing policy support of the environmentally harmful activities, in particular support of fossil fuels and wasteful agriculture, would reduce the environmental impacts of economic growth to one quarter of the present ones within one generation. The policies that abolish the entitlements for environmentally harmful activities also create fair competition and income distribution because of the reduction in rent-seeking behaviour.

13.2 Fostering Diversity of Sustainable Innovators

The diversity of sustainable innovations is vital for sustainable development. It is because prediction of the successful innovations is impossible and because all innovations have pros and cons, which cannot all be assessed and prevented beforehand. Cases illustrate the diversity of innovators and interactions between the innovators and various stakeholders. The innovators' cases are about the consumer innovators in solar power, tacit regional inventors in tourism, artists that create the natural blends, technology suppliers for sanitation and project developers of offices. The interactions are about consumers and corporations in the ethical consumption, sustainable investors and innovators, local energy initiatives that evolve into energy service companies and business development in renewable energy. These cases are presented in the ascending order of complexity.

13.2.1 Innovating Consumers

It is observed that the consumers, being users of products and services, exchange know-how about all kinds of products from relatively simple cooking recipes up to complicated house constructions. This know-how exchange, which evolves beyond the scope of policies and markets, generates innumerable improvements, nowadays labelled as the user innovations. The knowledge spillovers proceed usually among the peers, but if scaled up and disseminated, they generate businesses that develop markets. A sustainable innovation case is the solar boat competition when hundreds of students and professionals from all over the world come together for fun and experience though the solar power on leisure boats is also a potential global market of USD 28 billion. This competition has invoked technological changes in performance of boats whose rate, measured by speed, is higher than one found in the most dynamic businesses. Many novel solutions and products are developed and used, a few hundred existing businesses are involved, a dozen business start-ups are pursued, monetary benefits of this event outweighed its costs by three times and there is an educational spin-off. Two main impediments are observed. One is continuity.

Such events need organisation and funding and above all involvements of many stakeholders and volunteers. Second is slow adoption of novelties in the mature yacht- and ship-building industries. This case underpins that consumers, being user innovators, are a major source of innovations of simple and complex products and that social organisations enable to overcome the obstacles of costs and quality deficiencies and foster sustainable innovation markets. The policies in support of such social organisations generate sustainable innovations with high market potential along with large public appeal.

13.2.2 Tacit Inventors

Generating the tacit knowledge from within the region and from outside and strengthening the competitive position of the vested interests are two basic options for regional development. These options are referred to as the network policy and cluster policy. The case is about sustainable innovations for tourism development in a region where several million tourists a year constitute the largest regional business. The success of these policies is assessed by the number of inventions that turned into innovations and remained active after 10 years. In the network policy, the local inventors could get seed money for good proposals without formalities and the policymakers are made co-responsible for making good proposals. The success rate was exceptionally high: 52 % of the 73 granted inventions for sustainable tourism transport, arrangements and communication tools remained operating after 10 years. Social and tourism organisations were the most successful inventors. It indicates large tacit knowledge in the region, which can be used for innovations. When the granting process is structured in line with the European Union procedure, which means consortia of businesses and experts are organised to submit applications that are assessed by commissions of policymakers and experts, the success rate has dropped. In this cluster policy, 15 consortia with 64 enterprises and many experts are organised but two survived 10 years. The tacit innovators are more successful and less costly than the consortia of business and experts, and the policymakers with seed money are more helpful than consultants. Key factors in the regional development are scouting and fostering of the tacit knowledge and entrepreneurial skills and supporting their transformations into innovations. Nearly all regions can pay it. The issue is not about money but policy that encourages inventors though many proposals are unconventional and most of them fail. It needs political commitment and good facilitating organisation.

13.2.3 Arts Services

Environmental issues and qualities could be unrecognised because of poor metrics and imaginations, for instance, climate change was a non-issue 30 years ago and scarce darkness is still a non-issue. Artists interested in the environmental qualities could foster imaginations of the qualities and issues unrecognised so far, but art is considered a high-risk investment because of uncertain or intangible results. The arts services that combine the artistic and inventor's skills could create natural blends based on the unrecognised qualities. The possibilities of generating net income based on the arts services is experienced in 20 cities awarded as the European Capital of Cultures. All except one gained net income using the public–private funding. The arts services for the environmental qualities are envisaged for the European Capital of Culture 2018. A dozen of arts services are proposed, such as use of silence and darkness, experiences of spaciousness, links between culture and biodiversity, use of local dyes from plants and so on. Some already act. An economic assessment of an arts service shows that using environmental qualities of an island for the location theatres generates sufficient income for the high leverage of public funding and high rewards to artists. The policies that encourage the nexus of arts, environmental know-how and innovative skills into the arts services generate knowledge about the emerging environmental issues and about uses of environmental qualities for social benefit and generate income for artists. Public funds are needed because private investors consider the experimental arts as risky.

13.2.4 Technology Suppliers

When policies promote a specific technology, its suppliers' network usually evolves into lock-in, which impedes innovations. The increasing cost of performance is the consequence. The sanitation, which is an essential service but presently unavailable to nearly one third of the global population, illustrates such lock-in. The dominant suppliers' networks are interested in the elongation of sewage and adding technologies for wastewater treatment. The advanced elongation effectively reduces risks of human excrements but could be unattainable to billions of people in the next decades because the costs increase exponentially with more connections. The elongation network is driven by one policy criterion: maximise risk reduction at minimal cost. The maxi–min policy must change to reduce costs of sanitation. Three alternative technology networks have emerged, each one with pros and cons. The first one is the separation of wastewater streams at sources aiming to decrease losses and increase reuse. This reduces costs but needs behavioural changes. The second one is use of constructed wetlands for wastewater treatment under soil surface in vicinity of the pollution sources. This shortens sewage and avoids conventional wastewater treatment plants, but it needs much space. The third alternative is creation of value-adding services using wastewater, which can outweigh the costs but cannot meet all sanitary standards in the high-income countries. The policies that diversify the decision-making criteria in order to tune the technology suppliers' networks to the local capabilities and conditions induce innovations. Next to the maxi–min criterion that invoked elongation in sanitation, the loss prevention criterion can induce separation, the minimal capital criterion can induce the distributed sanitation system and the benefit criterion can induce the wastewater services. The policy that enables communities to decide about criteria tuned to their situations, given health and environmental aims, makes sanitation accessible to all people.

13.2.5 Office Alternatives

About half of the global jobs are in offices on distance from residential area. The distance grows because of the shifts on real estate markets. This generates commuting, entailing more congestion. The costs of commuting by car in the high-income countries approach 22 % of the average annual income. In addition, there are large nonmonetised social losses because communities are undermined and nature and landscapes are deplored. Congestion is often considered an issue of the transport policies, but better transport cannot stop its growth when work is bound to offices that are situated on larger distance from homes. Although most of the office work is done individually with use of information and communication technologies, nearly all work is done in the office buildings. The life cycle costs of four alternative office systems are assessed: the present mix of small and large offices, the highly concentrated offices, the distributed local offices for rent and office at home with extra space. All alternatives are socially beneficial compared to the present offices with commuting mainly because the concentrated offices can use public transport, and the distributed and home offices reduce commuting to nearly nil. The extra costs of the information and communication technologies and more space at home are outweighed. The life cycle costs of USD 23 billion per million employees of the present offices can be reduced by 15 % through the concentration, by 22 % due to the distribution and by 28 % for work at home. Lower costs of space and travels, and higher costs of the technologies do not change the key finding. Project developers could diversify the range of office work. The distributed offices are potentially profitable, add value to real estate and foster the community interactions. Large social benefits can be reached if policies stop supporting real estate development with entitlements, funding and tax exemptions for land use and infrastructure but accommodate social costs of congestion in the real estate prices.

13.2.6 Ethical Consumption

A conventional argument is that the sovereign consumers reveal their preference through purchases and this way determine qualities on markets. It is a contentious viewpoint. Practices are that many consumers state their preference for products with ethical attributes but not many reveal this in purchases. This gap between the stated and revealed preferences could be caused by the deficient product supplies. All products contain several functional qualities and ethical attributes compounded by subsequent suppliers in the product life cycle, but if there are trade-offs, consumers are confronted with intractable dilemma's in decision making. In addition, social mechanisms for rewarding altruistic behaviour are missing except in small groups of high morals. The suppliers do know what they compound, and the credible supplies are rewarded by premium market prices. Cases of life cycle management in the corporations considered as environmental frontrunners illustrate that firms innovate for the ethical consumption given the stated preferences. Corporate results are attained when new markets can be captured due to the ethical attributes, when innovations save costs in supply chain because materials can be used efficiently and when ethical attributes add value to consumers' products. Such supplies foster corporate credibility, entailing profits. Additional consumers' demands hardly invoke more ethical supplies and may even obstruct these if the demands are perceived risky for other products, but the innovating suppliers invoke ethical consumption. Instead of moralising about consumer behaviour, policies can enhance ethical supplies, for instance, through public procurement.

13.2.7 *Investors in Sustainable Innovations*

The global research and development expenditures increase faster than income along with even faster growth of inventions, still faster market introduction of innovations and even more the venture capital growth. The public funding and the risk-taking investors enable innovations. The research and development for inventions is supported with the public funds because the inventors are too risky for many private investors. The market introduction of innovations is driven by the risk-taking private investors in equity and supported with the public funding. Given the policies that prioritise sustainable development, one would expect that the sustainable innovators get much policy support, but it is not observed, for example, in the Netherlands. The public support of research and development on sustainable innovation is small, only 4 % of all support, though this share is larger than the share in private research and development investments. The public support of the market introduction of sustainable innovations is a smaller percentage compared to the support of all innovations, and it is much smaller than the private investments. Hence, competitors of the sustainable innovators get more public support. The reason for this policy inconsistency is not per se bad will or incapability of policymakers but information asymmetry between sustainable investors and innovators. Opposing opinions of the sustainable investors and innovators being the main interest groups constitute a policy risk because collisions between these interests can impede the policymaking. It is observed that the sustainable innovators and investors agree on many issues with the group but hardly between each other. Their opinions collide on three key issues. The innovators argue that the venture capital is scarce but the investors find that it is sufficient. The innovators advocate subsidies and the investors want policies that generate market demands for sustainable innovations. The investors advocate quality assurance and innovators oppose it because they confront costs. Resolving this information asymmetry would generate better policy, but bridging the interests needs cooperative models. The cooperative models would foster sustainable innovations because of the reduction in the investors' risks.

13.2.8 Energy Markets

Thousands of local energy initiatives emerged in the European Union. They pursue the distributed energy systems called smart grid. Some initiatives are social groups that advocate democratisation of the energy market, and many groups evolve into energy service companies. During 2008–2011, on annual average, about 3,600 energy firms have entered markets, which generated about 23,000 jobs. These companies constitute successful businesses during the period of economic crisis and stagnant energy consumption. The main markets cover substitution of fossil fuels for renewable energy, high-value-added products and services for residential electricity use and upgrading of the gas and heat reuse in businesses. The energy service companies are successful despite policy support of the vested fossil-fuel-based, large-scale corporations. The support covers tax exemptions and subsidies. The tax exemptions increase for the larger energy consumption. These tax exemptions are average of € 118 billion a year, which is 18 % of all energy costs. The tax exemptions also invoke price discounts for the large-scale energy consumption. This policy support reduces tax income, obstructs innovative start-ups, impedes progress in energy efficiency, undermines mitigation of climate change and encourages the rent-seeking behaviour instead of innovations. In addition, the policy support of fossil fuels is € 37 billion a year, excluding the support of nuclear energy. The largest support is allocated for oil resources, which counters the policy on energy independence. Until 2010, this policy support of fossil fuels was larger than the support of renewable energy. Abolishing tax exemptions and subsidies on energy production and consumption fosters clean smart growth.

13.2.9 Renewable Energy Business

The renewable energy business generates large income and contributes to better environmental qualities. It is assessed across the European countries what factors impede and drive this business. The main impediments are the limited space and large fossil fuel interests. The main drivers are the research and development and venture capital. These drivers can be strengthened. The basic policy options are support of the specific research and development and the generic risk reduction due to larger markets. Hundreds of policy instruments based on these options can be found. The policy support of the renewable energy business in the United States and in the European Union countries is compared. The United States support is larger, it is focused on the specific research and development, and its support of the fossil fuel interests is smaller compared to the European Union that is focused on the generic risk reduction through feed-in tariffs, which generates markets. The United States policy has generated a limited number of large, innovative firms; the European Union policy has generated many more new enterprises and twice as many jobs at the lower costs of policy support. The European Union policy is more cost-effective and socially beneficial. The feed-in tariffs policy has proven to be cost-effective. The regional policies can also reduce the investors' risks in the renewable energy business through cofinancing and regulations if politicians act across policy domains.

13.3 On Sustainable Development

The question was about how to pursue income growth and a better environment. In answering this question, two key messages are underpinned. One message is that the innovation-driven income growth and environmental qualities are positively interlinked because people demand good environment. They collide when innovations are suppressed because of the political oppression and entitlements given to the rent-seeking interests that harm environmental qualities because these obstruct innovations. The second message is that the diversity of innovators pursuing private and social interests is the driving factor for sustainable development. The diversity of sustainable innovations is fostered if policies enhance social capabilities to innovate through education, knowledge and stakeholder interactions and abolish perverse incentives and actions.

Index

A

Amenities, 36, 101
Arts, 2, 14, 21, 37, 41, 51, 61, 72–74, 77–85, 95, 164, 186–187

B

Backstop technologies, 6, 25–26
Behavioural, 1, 9, 10, 12, 13, 43, 84, 87, 91, 92, 97, 113, 114, 116, 117, 122, 125, 128–131, 148, 163, 184, 185, 187, 189, 190
Benefit, 1, 2, 7, 10, 11, 13–15, 36, 38, 41, 44, 49, 51, 52, 55, 57, 62, 65, 77, 79, 81, 85, 90, 95, 97, 107, 118, 125, 126, 133, 166, 168, 171–173, 185, 187–188
Bequest values, 38
Biodiversity, 2–4, 7, 21, 22, 24, 25, 30, 31, 38, 62, 83, 95, 96, 101, 183, 187
Business, 5–7, 9, 11–15, 22, 26, 29, 37–40, 43, 44, 49, 51, 54–56, 59–62, 65–69, 77, 81, 87, 88, 90, 96, 127, 128, 130, 132, 137–145, 147, 148, 150, 152–156, 158, 161–178, 185, 186, 190–191

C

Capital, 1–5, 8, 13, 20, 21, 29, 31, 39, 77–85, 89–91, 96, 97, 101, 103, 125–130, 132–135, 138, 139, 144, 161–164, 166–171, 173, 177, 187, 189, 190
Clusters, 14, 37, 59–62, 65–69, 168, 172, 186
Congestion, 14, 60, 99–103, 107, 188
Consumers, 3, 4, 13, 14, 29, 37, 40, 49–51, 57, 78, 79, 81, 85, 113–122, 127, 131, 137, 143, 144, 148, 163, 164, 168, 174, 185–186, 188, 189
Cultural attributes, 14, 39, 41–43
Culture, 4, 9, 14, 28, 36, 37, 41, 42, 62, 64, 66, 68, 72–74, 77–85, 187
Cycles, 5, 14, 20, 29, 91, 99–101, 103, 105–107, 113, 116–119, 121, 188–189

D

Decoupling, 6, 19, 21–31, 36, 183, 184
Demand-pull model, 8, 9, 126, 167
Distributed, 3, 5, 13, 51, 61, 87, 89, 92–95, 97, 99, 103, 107, 137, 138, 141–143, 148, 187–188, 190

E

Ecosystem services, 14, 35, 38, 39, 41, 43, 45, 78
Energy initiatives, 14, 137–139, 142, 143, 148, 172–173, 185, 190
Entrepreneurs, 3, 4, 8–10, 13, 31, 51, 59, 62, 65–69, 79, 84, 130–132, 162, 167, 169, 183, 186
Environmental Kuznets Curve, 22
Ethical attributes, 13, 14, 113–122, 138, 142, 188, 189
Ethical consumption, 14, 38–40, 43, 113–122, 184, 185, 188–189
Evolutionary, 1, 10, 11, 13, 49, 117, 162, 163
Existence values, 38, 41, 77–78

Export, 2, 23, 26, 27, 31, 56, 118
External effects, 6, 13, 38–39, 50, 60, 99–101, 106, 107, 116

F
Free riding, 12, 125
Functional qualities, 113–122, 138, 188

G
Greenhouse gasses, 3–4, 22, 24, 25, 30, 183
Gross domestic product (GDP), 2, 3, 19, 21–24, 27, 28, 30, 31, 35, 39–43, 45, 152, 164, 165, 175, 184

H
High income countries, 7, 14, 19, 22, 23, 25, 26, 28–31, 36, 37, 40, 44, 78, 90–92, 96, 100, 103, 107, 115, 116, 127, 128, 164, 183–184, 187

I
Induce, 14, 35, 42, 45, 88, 97, 99, 113, 143, 144, 168, 184, 187–188
Information and communication technologies (ICT), 5, 7, 23, 64, 67, 99, 100, 102, 104–107, 127, 139, 141, 188
Innovation process, 8–10, 125–126, 128, 129, 133–135, 167, 169
Innovation-rents, 8, 10, 19, 29–32, 35, 36, 184
IPAT model, 19–21, 23

K
Know-how, 1, 3, 9, 10, 12, 14, 19, 29–33, 42, 49, 50, 55–57, 59–62, 91, 97, 126, 132, 137, 138, 166, 172, 184, 185, 187
Knowledge spillovers, 10, 36, 59, 61, 69, 126, 131, 185
Knowledge work, 29, 35–37, 45, 50, 184

L
Labour, 1–3, 10, 19–21, 26, 28–32, 35, 36, 39, 60, 100, 103, 108, 118, 128, 184
Liabilities, 6, 23, 100, 101, 107, 168
Lock-in, 11, 14, 87, 88, 96, 97, 187
Low income countries, 19, 22–31, 36, 40, 44, 90, 93, 99, 116, 183–184

M
Market introduction, 8, 125–130, 135, 168, 169, 189

N
Natural blends, 35, 37–39, 77, 78, 83–85, 184, 185, 187
Neoclassic, 1, 2, 10–13, 113, 121, 162, 163
Networks, 9, 10, 13, 14, 23, 50, 59–65, 68, 69, 87, 88, 90–97, 99, 104, 108, 126, 133, 134, 138, 143, 162, 168, 169, 172, 178, 186, 187

O
Office work, 14, 99–108, 188
Option values, 38, 40
Organisation for Economic Co-operation and Development (OECD), 7, 23–28, 30, 37, 39, 40, 89, 116, 146, 147, 165, 170

P
Patents, 9, 10, 43, 45, 60, 127, 128, 184
Photovoltaic (PV), 49, 51–56, 172, 178
Policy support, 7, 14, 35, 43–45, 61, 63, 99, 125, 126, 128–130, 132, 133, 135, 146, 148, 149, 167, 169–171, 173, 184, 185, 189–191
Pollution, 2, 6, 10–14, 20–22, 24, 26, 35, 37–40, 43–45, 60, 88–90, 97, 99, 101, 107, 119, 138, 184, 187
Product life cycle, 29, 189

Q
Quasi-autonomous, 14

R
Renewable energy, 7, 14, 15, 23, 27, 37, 44, 61, 137, 138, 140–142, 145–148, 161–178, 185, 190–191
Research and development (R&D), 1, 9, 10, 49–51, 55, 60, 61, 125–131, 135, 161, 163, 166–170, 173, 177, 189–191
Risk-avoiding, 10, 12, 13, 88, 135, 169

S
Sanitation, 4, 14, 87–97, 109, 185, 187, 188
Scenarios, 6, 7, 162

Services, 3, 19, 35, 52, 61, 77, 87, 100, 113, 137, 173, 184
Skills, 1, 3, 9, 10, 31, 59, 61, 77, 78, 85, 116, 126, 131, 172, 184, 186, 187
Social costs, 2, 3, 6, 8, 11, 13, 43, 44, 62, 65, 88, 99–101, 103, 107, 171, 173, 183, 188
Spin-off, 11, 57, 67, 185
Stakeholders, 8, 11–13, 38, 56, 57, 59, 62, 63, 69, 70, 72–74, 84, 130, 132, 184–186, 191
Subsidies, 7, 35, 43–46, 51, 132–135, 137, 145–148, 161, 163–166, 168, 170, 172, 176, 184, 189, 190
Sustainable development, 1, 4, 7, 8, 30, 33, 35, 50, 78, 90, 126, 130, 132, 135, 183–185, 189, 191

T
Tacit, 4, 14, 50, 57, 59–74, 185, 186
Tax, 6, 7, 27, 28, 35, 42–46, 100–102, 126, 128, 143–149, 155, 156, 158, 172, 184, 188, 190

Technological change, 3, 5, 12, 14, 19, 26, 29–31, 45, 46, 55, 57, 184, 185
Tourism, 4, 14, 37, 38, 42, 52, 59, 61–74, 80, 84, 89, 185, 186
Transition, 8, 11, 12, 52
Transport, 4, 5, 22, 28, 37, 40, 59, 61, 63–66, 68, 70, 87, 90, 92, 96, 99–107, 109, 110, 118, 140, 141, 145, 146, 150, 171–173, 178, 186, 188
Trial and error, 50, 51, 126

U
User innovations, 14, 49–56, 185
User values, 38
Utensils, 29, 50, 79

W
Water, 2, 21, 37, 54, 61, 79, 89, 118, 183
Welfare, 1–3, 8, 42–44, 51, 57, 77, 99, 117, 183, 191
Willingness to pay, 36, 38, 79, 83, 115, 116, 131

MIX
Papier aus verantwortungsvollen Quellen
Paper from responsible sources
FSC® C105338

If you have any concerns about our products, you can contact us on
ProductSafety@springernature.com

In case Publisher is established outside the EU, the EU authorized representative is:
**Springer Nature Customer Service Center GmbH
Europaplatz 3, 69115 Heidelberg, Germany**

Printed by Libri Plureos GmbH
in Hamburg, Germany